ADVANCED GEODYNAMICS

David Sandwell developed this advanced textbook over a period of nearly 30 years for his graduate course at Scripps Institution of Oceanography. The book augments the classic textbook *Geodynamics* by Don Turcotte and Jerry Schubert, presenting more complex and foundational mathematical methods and approaches to geodynamics. The main new tool developed in the book is the multi-dimensional Fourier transform for solving linear partial differential equations. The book comprises nineteen chapters, including: the latest global data sets; quantitative plate tectonics; plate driving forces associated with lithospheric heat transfer and subduction; the physics of the earthquake cycle; postglacial rebound; and six chapters on gravity field development and interpretation. Each chapter has a set of student exercises that make use of the higher-level mathematical and numerical methods developed in the book. Solutions to the exercises are available online for course instructors, on request.

DAVID T. SANDWELL is a Professor of Geophysics at Scripps Institution of Oceanography, University of California San Diego. His research is focused on marine tectonics and geodynamics, and he develops global marine gravity and topography models from satellite altimetry and ship soundings. He is author of more than 190 research papers. He is a Fellow of Geological Society of America, the American Geophysical Union, the American Association for the Advancement of Science, and a member of the U.S. National Academy of Sciences

'*Advanced Geodynamics* brings the unique perspective of a leading geophysicist to the solution of a wide array of problems in geodynamics. The approach emphasizes the use of advanced mathematics, in particular the Fourier transform, to obtain a quantitative understanding of the processes involved in shaping the Earth's surface. The advanced mathematical approach not only enhances the elegance of the solutions, but it enables the consideration of many problems not accessible with less sophisticated mathematical methods. The choice of problems benefits from the deep physical insights of the author to their solutions. The book discusses the physical processes involved in plate tectonics and the earthquake cycle and provides the latest relevant observational data sets. An emphasis is also placed on the use of gravity data to learn about these processes. The book is the product of decades of teaching by the author and is a must read for students of the physics of the Earth with the appropriate mathematical background.'

Gerald Schubert,
University of California, Los Angeles; co-author of *Geodynamics*

'Most authors would find writing a sequel to Turcotte and Schubert's classic book on *Geodynamics* a daunting task. Not so for David Sandwell, whose first book is a wonderful mix of observations and theory, elegant mathematics and a focus on the oceans and the Fourier method which together help illuminate some of the fundamental physical processes that underlie plate tectonics.'

Tony Watts,
University of Oxford; author of Isostasy and Flexure of the Lithosphere

'*Advanced Geodynamics: The Fourier Transform Method* by David Sandwell is a godsend for advanced undergraduate students, graduate students, and researchers actively engaged in the broad area of geodynamics. It complements the classic *Geodynamics* book by Turcotte & Schubert in a way nothing else could: by elevating the treatment to real, cutting-edge research problems via Fourier transforms that deliver simple and elegant solutions to complicated science problems.'

Paul Wessel,
University of Hawaii

ADVANCED GEODYNAMICS

Fourier Transform Methods

DAVID T. SANDWELL
University of California, San Diego

CAMBRIDGE
UNIVERSITY PRESS

CAMBRIDGE
UNIVERSITY PRESS

University Printing House, Cambridge CB2 8BS, United Kingdom

One Liberty Plaza, 20th Floor, New York, NY 10006, USA

477 Williamstown Road, Port Melbourne, VIC 3207, Australia

314–321, 3rd Floor, Plot 3, Splendor Forum, Jasola District Centre, New Delhi – 110025, India

103 Penang Road, #05–06/07, Visioncrest Commercial, Singapore 238467

Cambridge University Press is part of the University of Cambridge.

It furthers the University's mission by disseminating knowledge in the pursuit of education, learning, and research at the highest international levels of excellence.

www.cambridge.org
Information on this title: www.cambridge.org/9781316519622
DOI: 10.1017/9781009024822

First published 2022

Printed in the United Kingdom by TJ Books Limited, Padstow Cornwall

A catalogue record for this publication is available from the British Library.

Library of Congress Catalo]-ging-in-Publication Data
Names: Sandwell, David T., author.
Title: Advanced geodynamics : the fourier transform method /
David T. Sandwell.
Description: New York : Cambridge University Press, 2021. |
Includes bibliographical references and index. |
Identifiers: LCCN 2021028137 (print) | LCCN 2021028138 (ebook) |
ISBN 9781316519622 (hardback) | ISBN 9781009024822 (epub)
Subjects: LCSH: Geodynamics. | BISAC: SCIENCE / Earth Sciences / Geology
Classification: LCC QE501 .S242 2021 (print) | LCC QE501 (ebook) |
DDC 551.1–dc23
LC record available at https://lccn.loc.gov/2021028137
LC ebook record available at https://lccn.loc.gov/2021028138

ISBN 978-1-316-51962-2 Hardback

To Susan, Katie, Melissa, Nick, and Cassie

Eddie Would Go

Contents

Colour plates can be found between pages 150 and 151.

Introduction

Geodynamics by Turcotte and Schubert (2014) provides a deterministic, physics-based exposition of solid-Earth processes at a mathematical level assessable to most students. This classic textbook begins with a clear and concise overview of plate tectonics, followed by stress and strain in solids, elasticity and flexure, heat transfer, gravity, fluid mechanics, rock rheology, faulting, flows in porous media, and chemical geodynamics; the latest edition has sections on numerical modeling. I have used this textbook, including earlier editions, in a graduate level class for the past 28 years to prepare students in quantitative modeling of Earth processes. The book uses a minimum of mathematical complexity, so it can be understood by a wide range of students in a variety of fields. However, this more limited mathematical approach does not provide the graduate student with the tools to develop more advanced models having three-dimensional geometries and time dependence.

This new book, *Advanced Geodynamics*, was developed to augment *Geodynamics* with more complex and foundational mathematical methods and approaches. The main new tool is multi-dimensional Fourier analysis for solving linear partial differential equations. Each chapter has a set of homework problems that make use of the higher-level mathematical and numerical methods. These are intended to augment the already excellent homework problems provided in *Geodynamics*. Detailed solutions are available from the author on request.

Chapter 1 – Observations Related to Plate Tectonics
This chapter reviews the global observations that were used to develop and refine the theory of plate tectonics. These include the latest maps of topography, marine gravity, seismicity, seafloor age, crustal thickness, and lithospheric thickness.

1

This chapter also provides the global grids as overlays to Google Earth for exploration and interaction by students. In addition, all the data and tools needed to prepare the global maps using Generic Mapping Tools (GMT) are provided at the Cambridge web site.

Chapter 2 – Fourier Transform Methods in Geophysics
This chapter provides a brief overview of Fourier transforms and their properties including: similarity, shift, derivative, and convolution as well as the Cauchy residue theorem for calculating inverse transforms. These tools are used throughout the book to solve multi-dimensional linear partial differential equations (PDE). Some examples include: Poisson's equation for problems in gravity and magnetics; the biharmonic equation for problems in linear viscoelasticity, flexure, and post-glacial rebound; and the diffusion equation for problems in heat conduction. There are two approaches to solving this class of problem. In some cases, one can derive a fully analytic solution, or Green's function, to the point-source problem. Then a more general model is constructed by convolution using the actual distribution of sources. We focus on the second semi-analytic approach since it can be used to solve more complicated problems where the development of a fully analytic Green's function is impossible. This involves using the derivative property of the Fourier transform to reduce the PDE and boundary conditions to algebraic equations that can be solved in the transform domain. A more general model can then be constructed by taking the Fourier transform of the source, multiplying by the transform domain solution, and performing the inverse transform numerically. When dealing with spatially complex models, the second approach can be orders of magnitude more computationally efficient, because of the efficiency of the fast Fourier transform algorithm.

Chapter 3 – Plate Kinematics
This chapter is focused on the basics of plate kinematics and relative plate motions. Students are encouraged to learn the names of the major plates, the plate boundaries, and triple junctions. We then review the rules governing the relative motions across the three types of plate boundaries, spreading ridges, transform faults, and subduction zones and use these rules for triple junction closure of the relative velocity vectors. The remainder of the chapter is concerned with plate motions on a sphere using vector calculus. The exercises involve calculations of plate motions and plate circuit closure using published rotation poles.

Chapter 4 – Marine Magnetic Anomalies
This chapter uses the Fourier transform tools developed in Chapter 2 to compute the scalar magnetic field that is recorded by a magnetometer towed behind a ship,

given a magnetic timescale, a spreading rate, and a skewness. We first review the origin of natural remnant magnetism, to illustrate that the magnetized layer is thin compared with its horizontal dimension. Then the relevant differential equations are developed and solved under the ideal case of seafloor spreading at the north magnetic pole. Anomalies that formed at lower latitudes have a skewness that causes a wavelength-dependent phase shift. The exercises include the calculation of the magnetic anomalies associated with seafloor magnetic stripes and comparisons with shipboard magnetic data to establish the seafloor spreading rate and skewness.

Chapter 5 – Cooling of the Oceanic Lithosphere
This chapter uses the Fourier transform method to solve for the temperature in the cooling oceanic lithosphere for half space and plate cooling models. For researchers in the areas of marine geology and geophysics, this is the essence of geodynamics since it explains the age variations of marine heat flow, seafloor depth, elastic thickness, and geoid height. The cooling models are also used to calculate the driving forces of plate motions including ridge push and slab pull. We focus on the buoyancy of the lithosphere as a function of crustal thickness to explain the conditions when subduction is possible. A highlight of this chapter are nine challenging heat flow exercises based on publications including thermal evolution of an oceanic fracture zone, lithospheric reheating from a mantle plume, and frictional heating during an earthquake.

Chapter 6 – A Brief Review of Elasticity
This chapter reviews stress, strain, and elasticity in three dimensions using tensors. There is a brief presentation of tensor rotations, principal stress, and stress invariants. The principal stress vs. strain is inverted using the symbolic algebra in MATLAB. This is used to translate the Lamé elastic constants to Poisson's ratio and Young's modulus. The plane stress formulation is used to develop the moment versus curvature relationship for a thin elastic plate. This chapter is intended as a review and reminds some students that they need to master this material.

Chapter 7 – Crustal Structure, Isostasy, Swell Push Force, and Rheology
This chapter covers four topics. The first is the basic structure of the oceanic and continental crust. The second and third topics are the vertical and horizontal force balances due to variations in crustal thickness. The vertical force balance, isostasy, provides a remarkably accurate description of variations in crustal thickness based on a knowledge of the topography. The horizontal force balance provides a lower bound on the force needed to maintain topographic variations on the Earth. The fourth topic is the rheology of the lithosphere. How does the lithosphere strain in response to applied deviatoric stress? The uppermost part of the lithosphere is

cold, so frictional sliding along optimally oriented, pre-existing faults governs the strength. At greater depth, the rocks can yield by non-linear flow mechanisms. The overall strength-versus-depth profile is called the yield-strength envelope (YSE). The integrated yield strength transmits the global plate tectonic stress. Moreover, the driving forces of plate tectonics cannot exceed the integrated lithospheric strength. This provides an important constraint on the geodynamics of oceans and continents.

Chapter 8 – Flexure of the Lithosphere
This chapter covers lithospheric flexure theory for an arbitrary vertical load. The approach is similar to the solutions of the marine magnetic anomaly problem, the lithospheric heat conduction problem, the strike-slip fault problem, and the flat-Earth gravity problem. In all these cases, we use the Cauchy residue theorem to perform the inverse Fourier transform. In a later chapter we combine this flexure solution with the gravity solution to develop the gravity-to-topography transfer function. Moreover, one can take this approach further, to develop a Green's function relating temperature, heat flow, topography, and gravity to a point heat source. In addition to the constant flexural rigidity solution found in the literature, we develop an iterative solution to flexure with spatially variable rigidity.

Chapter 9 – Flexure Examples
This chapter provides practical examples of flexural models applied to structures in the lithospheres of Earth and Venus. The models are all solutions to the thin and thick-plate flexure equation, with a variety of surface loads, sub-surface loads, and boundary conditions. Both gravity and topography data are used to constrain the models. We provide a numerical example that takes arbitrary topography and gravity anywhere on the Earth and uses Generic Mapping Tools to find the best elastic thickness and densities. A unique feature of this chapter is a comprehensive discussion of the non-linear relationship between plate bending moment and curvature that dominates at all subduction zones. This chapter includes eight challenging flexure exercises based on publications including: ice shelf flexure, seamount flexure, fracture zone flexure, trench and outer rise yield strength and fracturing, and flexure on Venus.

Chapter 10 – Elastic Solutions for Strike-Slip Faulting
This chapter provides the mathematical development for the deformation and strain pattern due to a strike-slip fault in an elastic half space. We develop the solution from first principles using the Fourier transform approach. This approach

does not explicitly use dislocations but simulates dislocations using body force couples following Steketee (1958) and Burridge and Knopoff (1964). The main advantage of this method is that it is easily extended to three dimensions as well as complicated fault geometries. We also demonstrate the inherent non-uniqueness of inverting for slip versus depth from surface geodetic data yet show that the overall seismic moment is well resolved by surface data. The exercises at the end of the chapter illustrate the use of the 3-D Fourier transform, the Cauchy residue theorem, and computer algebra to solve for the response of an elastic half space to 3-D vector body forces.

Chapter 11 – Heat Flow Paradox
This chapter is a quantitative investigation of the heat flow paradox that relates the expected frictional heating on a fault to the measurements of surface heat flow above the fault (e.g., (Lachenbruch and Sass, 1980)). A straightforward calculation, using a reasonable coefficient of friction, predicts measurably high heat flow above the fault that is not observed. We also investigate the maximum tectonic moment that could be sustained by a fault and show that it is at least an order of magnitude greater than what is observed. Finally, we discuss the implications in terms of fault strength and earthquake predictability.

Chapters 12 – The Gravity Field of the Earth, Part 1
This chapter provides a brief introduction to physical geodesy that describes the size and shape of the Earth and its gravity field. We decompose the Earth's reference gravity field into a spherical term and terms related to hydrostatic flattening by rotation. Superimposed on this reference model are anomalies discussed in later chapters. This chapter also describes how the reference Earth model has been developed and defined using satellite observations.

Chapters 13 – Reference Earth Model: WGS84
This chapter is a summary of the reference shape and gravity field of the Earth as defined by the WGS84 parameters. Deviations from this reference model are defined in terms of geoid height, gravity anomaly, and deflections of the vertical.

Chapter 14 – Laplace's Equation on Spherical Coordinates
This chapter introduces spherical harmonics and their properties for representing planetary gravity fields. We explain how the harmonic decomposition of a function on a sphere is analogous to the Fourier series decomposition of a 2-D function in Cartesian coordinates. We then use this spherical harmonic formulation to solve

Laplace's equation and discuss upward continuation. Finally, we describe how the Earth's gravity field is represented as spherical harmonic coefficients and their time variation.

Chapter 15 – Laplace's Equation in Cartesian Coordinates and Satellite Altimetry
This chapter is focused on shorter wavelength components of the gravity field that are best represented in Cartesian coordinates using Fourier series. The Fourier transform of Laplace's equation is used to illustrate upward continuation as well as the connection between the anomalous potential (geoid height), its first derivatives (gravity anomaly and deflections of the vertical), and the second derivatives (gravity gradient tensor). This chapter also contains a quite complete discussion of satellite radar altimetry and how it is used to recover short wavelength variations in gravity which provides an important tool for investigating plate tectonics.

Chapter 16 – Poisson's Equation in Cartesian Coordinates
This chapter is focused on solving Poisson's equation using Fourier transforms. This solution is used to generate models of the disturbing potential and its derivatives from a 3-D density model. One approach is to perform a convolution of the Green's function with the 3-D density model. However, this approach, which appears in most textbooks, is error prone, computationally inefficient, and almost never used in modern publications. Instead, we illustrate the Fourier transform approach where the model is divided into layers and the density of each layer is Fourier transformed, upward continued, and summed to generate a surface model. A uniform density leads to the Bouguer slab correction. Finally, we develop Parker's exact formula for computing the gravity model for a layer with non-uniform topography.

Chapter 17 – Gravity/Topography Transfer Function and Isostatic Geoid Anomalies
This chapter combines thin-elastic plate flexure theory with the solution to Poisson's equation, to develop a linear relationship between gravity and topography. We discuss three uses of this relationship: (1) If both the topography and gravity are measured over an area that is several times greater than the flexural wavelength, then the gravity/topography relationship (in the wavenumber domain) can be used to estimate the elastic thickness of the lithosphere and/or the crustal thickness. (2) At wavelengths greater than the flexural wavelength, where features are isostatically compensated, the geoid/topography ratio can be used to estimate the depth of compensation of crustal plateaus and hot-spot swells. (3) If the gravity

field is known over a large area, but there is rather sparse ship-track coverage, the topography/gravity transfer function can be used to interpolate the seafloor depth between the sparse ship soundings. Finally, we show that the geoid height for isostatically compensated topography is proportional to the swell push or ridge push force so under ideal conditions, one component of the plate driving force can be measured from the geoid height.

Chapter 18 – Postglacial Rebound

This chapter considers the classic postglacial rebound problem using the Fourier transform approach. The chapter relies on Chapter 6 in *Geodynamics* where the differential equations for viscous flow of an incompressible fluid are developed. We then solve for the response of a viscous half space to an arbitrary initial topography to illustrate the effect of load wavelength on the relaxation time. In addition, an elastic lithosphere is added to the viscous half space to simulate the present-day collapse of the flexural forebulge on the perimeter of the major Laurentide and Fennoscandia ice loads.

Chapter 19 – Driving Forces of Plate Tectonics

This chapter discusses the three major driving forces of plate motion — ridge push, slab pull, and viscous drag. In previous chapters we showed that the ridge-push force is proportional to the age of the cooling ocean lithosphere. In this chapter we focus on the slab pull force which depends on age of the subducted lithosphere as well as the depth that the slab remains coupled to the surface. We use results from recent publications to calculate the positive and negative buoyancy of three major phase changes. (1) The basalt crust and depleted layer undergo a phase change to the higher density eclogite. (2) Endothermic phase changes (positive Clapeyron slope) produce a zone of increased density in the cold lithosphere for a large part of the transition zone between depths of 310 and 660 km. (3) Exothermic phase changes (negative Clapeyron slope) below 660 km result in a zone of decreased density between depths of 660 and 720 km. Finally, we discuss the magnitudes of the forces for subduction of a small young plate as well as a large old plate to illustrate that the slab pull force (thermal plus phase changes) dominates ridge push and the difference must be attributed to the drag force.

The mathematical developments refer back to the section or equation in *Geodynamics* where the solutions are provided. Note that *Geodynamics* contains much more information than is provided in this new book *Advanced Geodynamics*, so both will be needed for a graduate-level geodynamics course.

Acknowledgments   I thank Jerry Schubert, Tony Watts, Paul Wessel, numerous students for reviewing and commenting on the manuscript. The Generic Mapping Tools (GMT) (Wessel et al., 2019) were used extensively in data analysis and to generate figures. The individual chapters in this book were first developed as lecture notes and converted to LaTeX by the people at Dangerous Curve (typesetting@dangerouscurve.org), some of whom have been doing typography since 1979.

1

Observations Related to Plate Tectonics

1.1 Global Maps

The plate tectonic model states that the outer shell (lithosphere) of the Earth is divided into a small number of nearly rigid plates which slide over the weak asthenosphere. The plates are the surface thermal boundary layer (TBL) of mantle convection, and descending slabs are the primary active components of the convective system. Plate boundaries are generally narrow and are characterized by earthquakes and volcanoes.

It is useful to assess the global data sets that are most relevant to plate tectonics. Below are a series of global maps that help to confirm various aspects of plate tectonic theory. Plate boundaries are classified as ridges, transform faults, or subduction zones based on basic observations of topography (Figures 1.1 and 1.2), gravity anomaly (Figure 1.3), and seismicity (Figure 1.4). Remarkably, the axes of nearly all seafloor spreading ridges lie at a depth of 2500–3000 m below sea level, which is the level of isostasy for a hot thin lithosphere. Depths gradually increase away from the ridges, because of cooling and thermal contraction, so old ocean basins are commonly 4500–5000 m deep. Fracture zones and aseismic ridges also show up on these maps. Global seismicity (magnitude >5.1, Figure 1.4) highlights the plate boundaries and reveals their tectonic style. Shallow normal-faulting earthquakes (<30 km deep) are common along slow-spreading ridges, but largely absent along faster-spreading ridges where the plates are too thin and weak to retain sufficient elastic energy to generate large earthquakes. Transform faults are characterized by relatively shallow (<30 km) strike-slip earthquakes, and they are common along both fast, and slow-spreading ridges. The deeper-earthquakes (green and blue dots in Figure 1.4) occur only in subduction zones where sheets of seismicity (i.e., Wadati-Benioff zone) are critical evidence that the relatively cold lithosphere is subducting back into the mantle. But even convergent boundaries are characterized by shallow extensional earthquakes on the ocean side of the

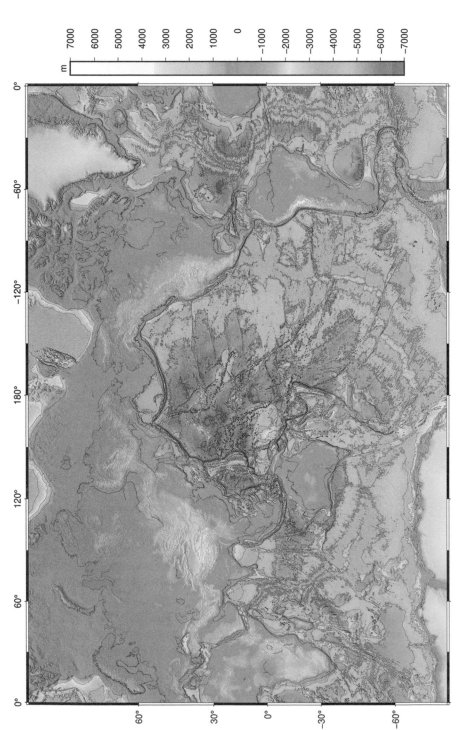

Figure 1.1 Topography of the Earth based on a global compilation of land data (SRTM and other sources) and ocean data from a combination of sparse ship soundings and marine gravity anomalies derived from satellite altimetry (Smith and Sandwell, 1997; Tozer et al., 2019). (For a color version of this figure, please see the color plate section.)

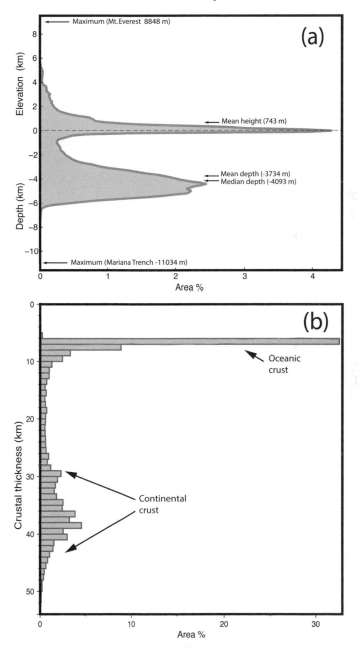

Figure 1.2 (a) Histogram of topography (100 m bins), based on a global compilation (Becker et al., 2009). The Earth has a bimodal histogram with the largest peak representing the land and submerged continental shelf. The second largest peak represents the median ocean depth of about 4000 m. This ocean peak has two subpeaks of unknown origin, but perhaps representing the relatively uniform depth of hot-spot swells. (b) Histogram of crustal thickness (1 km bins) based on refraction seismology as well as receiver function analyses (Laske et al., 2013). Crustal thickness also has a bimodal distribution where oceanic crust is 6–7 km thick and continental crust is usually greater than 28 km thick.

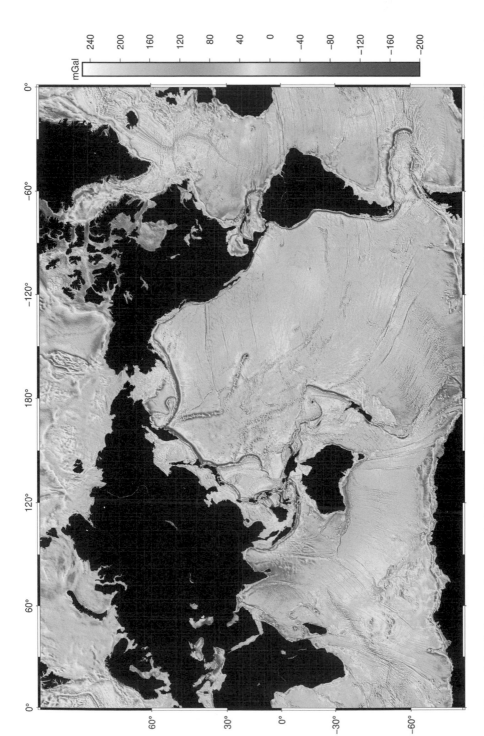

Figure 1.3 Marine gravity anomaly based on satellite altimetry (Sandwell et al., 2014). (For a color version of this figure, please see the color plate section.)

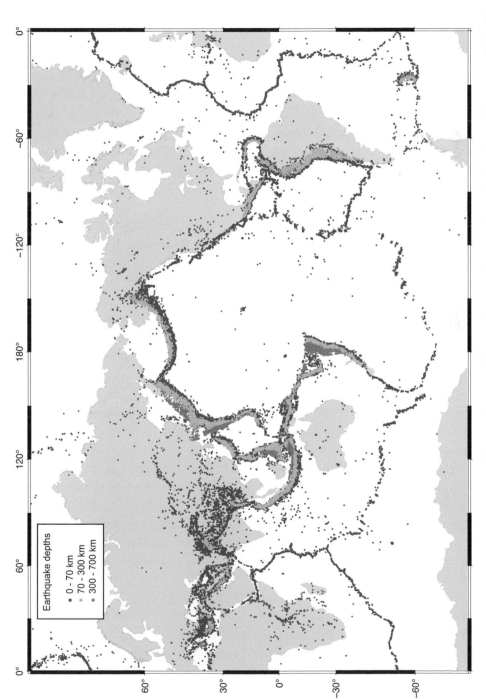

Figure 1.4 Well-located earthquakes with magnitude >5.1 reveal the global plate boundaries (Engdahl et al., 1998). Shallow earthquakes (0–70 km: red) provide clear definition of the boundaries of the oceanic plates, but have a more diffuse distribution on the continents. Intermediate (70–300 km: green) and deep (300–700 km: blue) earthquakes occur in subduction zones and are the primary evidence for lithospheric subduction to at least 670 km. (For a color version of this figure, please see the color plate section.)

trenches. Some regions (e.g., Africa, Asia, western North America, and the Indian Ocean) have distributed earthquake activity indicating broad deformational zones. Topography and seismicity provide strong evidence for tectonic activity but little or no information on the rate of plate motion.

Marine magnetic anomalies, combined with relative plate motion directions based on satellite altimeter measurements of fracture-zone trends, have been used to construct a global age map (Figure 1.5) of the relatively young (<180 Myr) oceanic lithosphere. Finally, the distribution of off-ridge volcanoes that have been active during the Quaternary mainly occur directly behind trenches where wet subducting slabs reach asthenospheric depths and trigger back-arc volcanism (Figure 1.6). A few active volcanoes occur within the interiors of the plates and in diffuse extensional plate boundaries.

The geoid (Figure 1.7) shows little correlation at long wavelengths with surface tectonics, and primarily reflects mass anomalies deep in the mantle. It is expected that the dynamic topography–the topography not due to crustal and near-surface variations–and the stress-state of the lithosphere will also reflect deep density differences. Insofar as volcanoes correlate with high surface elevations and extensional stress, one expects correlation of volcanoes with deep mantle structure, even if there is no material transfer. These maps are available for viewing in Google Earth at the following site: `topex.ucsd.edu/geodynamics/tectonics.kmz`.

A global map of crustal thickness (Figures 1.8 and 1.2), based on refraction seismology as well as receiver function analyses, shows the major contrast between oceanic and continental crust. Ocean crustal thickness is relatively uniform (6–7 km). In contrast, the crustal thickness under the continents is generally 30–40 km in areas where the topography is within a few hundred meters of sea level. Areas of high elevation such as the Andes Mountains, the Himalayas, and the Tibetan Plateau have much thicker crust. As discussed throughout the text, the lack of high-amplitude, long-wavelength gravity anomalies is evidence that these large-scale topographic variations are isostatically compensated—mostly by variations in crustal thickness.

The final map shows lithospheric thickness derived from surface wave tomography (Figure 1.9). Over the ocean, the thickness of the lithosphere increases with the age of the plates. The thickest oceanic lithosphere occurs along the western side of the

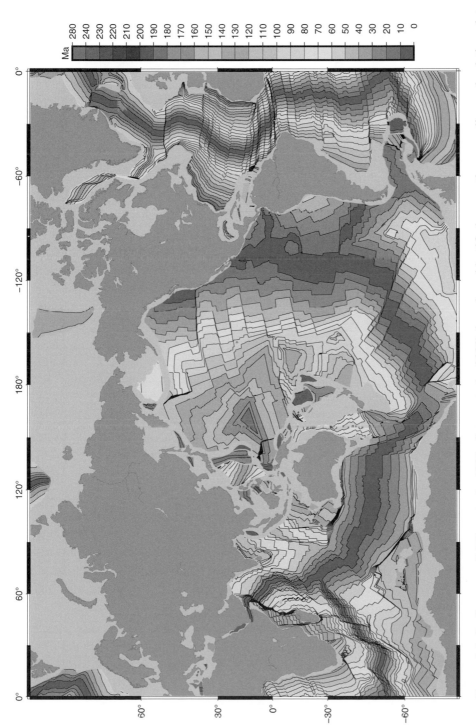

Figure 1.5 Seafloor age (Müller et al., 1997; Seton et al., 2020) based on identified magnetic anomalies and relative plate reconstructions along trends identified in satellite altimeter measurements of marine gravity. Ages in the Cretaceous quiet zone (64–127 Ma), the Jurassic (145–200 Ma) and older have poor control. (For a color version of this figure, please see the color plate section.)

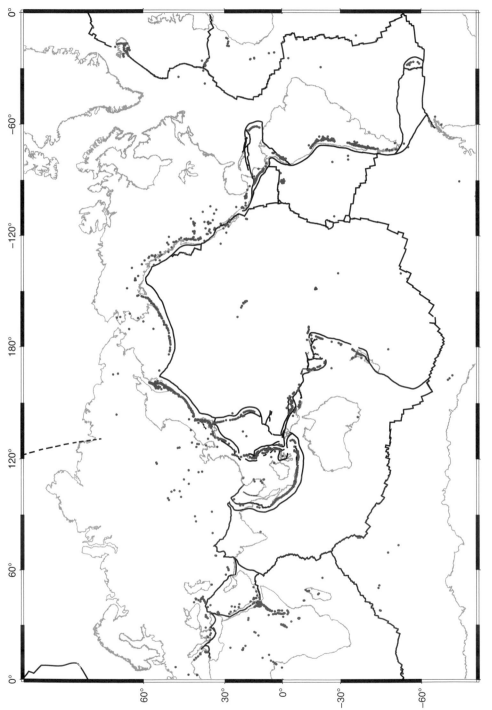

Figure 1.6 Plates, plate boundaries, and subaerial Quaternary volcanoes. The dozen or so tectonic plates are separated by spreading centers, transform faults, and subduction/thrust faults. The majority of active or recently active volcanoes (Siebert and Simkin, 2002) are associated with convergent plate boundaries. (For a color version of this figure, please see the color plate section.)

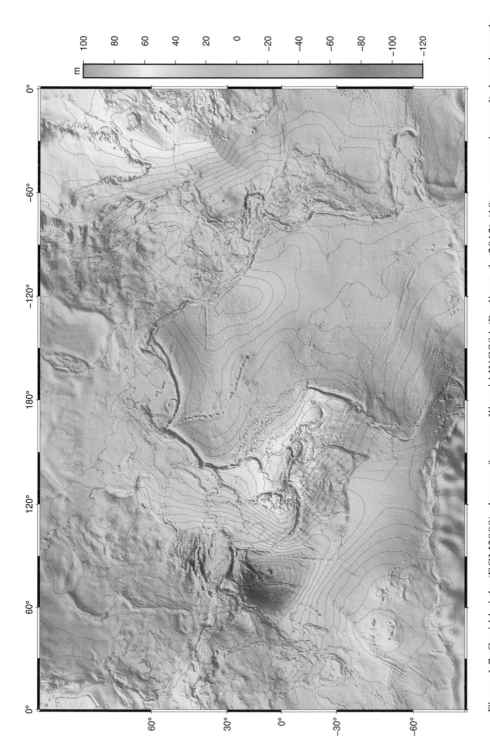

Figure 1.7 Geoid height (EGM2008) above reference ellipsoid WGS84 (Pavlis et al., 2012) (10 m contour interval), based mostly on satellite tracking data and some terrestrial gravity anomaly measurements. Unlike topography, seismicity, and age shown in the other maps, the geoid is poorly correlated with surface tectonics, except in areas where mature lithosphere has subducted in the western Pacific. (For a color version of this figure, please see the color plate section.)

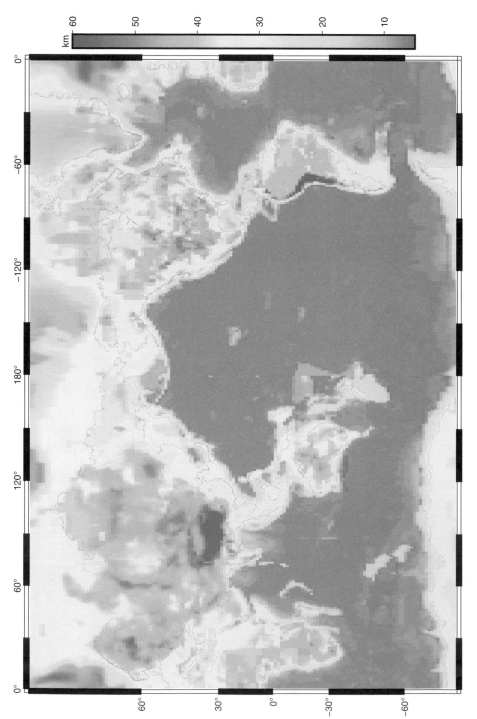

Figure 1.8 Crustal thickness (1 degree averages) based on refraction seismology as well as receiver function analyses (Laske et al., 2013). Gravity anomalies were used to estimate crustal thickness in areas with no constraints. (For a color version of this figure, please see the color plate section.)

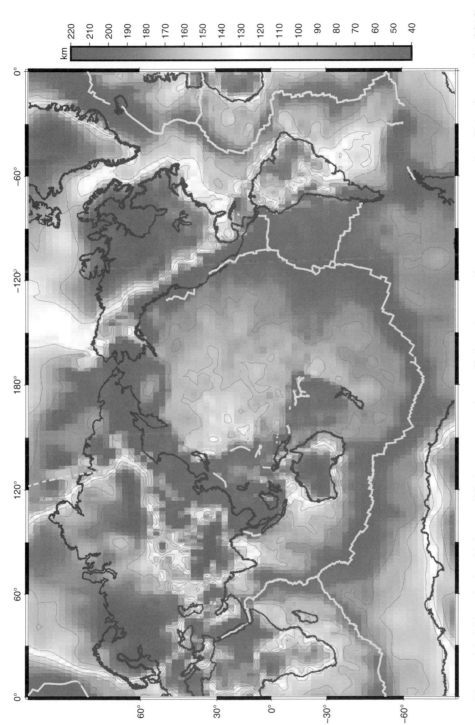

Figure 1.9 Lithospheric thickness (20 km contour interval) based on surface wave tomography (Priestley et al., 2018). Present-day spreading ridges are shown as yellow line. (For a color version of this figure, please see the color plate section.)

North and South Atlantic basins as well as the Western Pacific. The lithosphere is very thick (>150 km) beneath the continental cratons of South Africa, Australia, northern Eurasia, northeastern North America, eastern South America, Greenland, and East Antarctica. Thinner continental lithosphere occurs in tectonically active areas. One prominent exception is Tibet and the mountain ranges of central Asia where there is active crustal shortening and underthrusting of the Indian plate beneath Tibet.

1.2 Exercises

Exercise 1.1 Install Google Earth on your computer and download the global tectonic maps (`topex.ucsd.edu/geodynamics/tectonics.kmz`). Identify the following triple junctions, and use the overlays to determine the type of deformation (R-ridge, F-transform fault, or T-trench) for each of the three boundaries. Do this for the Mendocino, Galapagos, Chile, Bouvet, Azores, and Indian Ocean triple junctions.

Exercise 1.2 Sketch a topographic profile across the Atlantic Ocean following a tectonic flow line (i.e., perpendicular to isochrons). The profile should extend from the east coast of North America to the west coast of Africa. Label the major topographic features. Provide approximate depths for the major topographic features. Sketch a second profile that extends from the ridge axis to the coast of North America and also intersects the Island of Bermuda. What are some major differences between this profile and the first profile?

Exercise 1.3 Where is the youngest ocean floor? Where is the oldest ocean floor? What are their approximate ages?

Exercise 1.4 What types of earthquake focal mechanism occur on the three main types of plate boundary?

Exercise 1.5 Use the book *Geodynamics* (Turcotte and Schubert, 2014) to complete Table 1.1. Devise a thought experiment to measure each quantity. The experiments should be physically realistic, but not necessarily practical.

Example temperature: One could use a thermometer to measure temperature, but that depends on knowing the coefficient of thermal expansion. One could use the definitions of the freezing/boiling point of water to define temperatures of 0 °C and 100 °C. Or one could use the Stefan–Boltzmann law to measure temperature by measuring radiation L from a black body at temperature T.

$$L = \sigma T^4 \tag{1.1}$$

where σ is the Stefan–Boltzmann constant (5.67×10^{-8} W m^{-2} K^{-4}).

Table 1.1.

Parameter	Symbol	Units	Typical value and quantity
Temperature	*T*	°C or K	100 °C is the boiling point of water at 1 atm. of pressure.
Thermal conductivity			
Heat capacity			
Density			
Coefficient of thermal expansion (volumetric)			
Acceleration of gravity			
Gravitational constant			
Young's modulus			
Poisson's ratio			
Shear modulus			
Bulk modulus			
Dynamic viscosity			

Acknowledgments: The data provided in these figures represent decades of data collection by thousands of scientists. Figures were constructed using Generic Mapping Tools (GMT) (Wessel and Smith, 1995; Wessel et al., 2019). All the global maps can be reconstructed or customized using the data and GMT scripts at the following site: topex.ucsd.edu/geodynamics/global_maps.tgz.

2

Fourier Transform Methods in Geophysics

2.1 Introduction

Fourier transforms are used in many areas of geophysics such as image processing, time series analysis, and antenna design. Here we focus on the use of Fourier transforms for solving linear partial differential equations (PDE). Some examples include Poisson's equation for problems in gravity and magnetics; the biharmonic equation for problems in linear viscoelasticity; and the diffusion equation for problems in heat conduction. We do not treat the wave equation in this book, because there are already many excellent books on seismology. For each of these problems, we search for the Green's function that represents the response of the model to a point source. There are two approaches to solving this class of problem. In some cases, one can derive a fully analytic solution, or Green's function, to the point-source problem. Then a more general model can be constructed by convolving the actual distribution of sources with the Green's function. A familiar example is the case of constructing a gravity anomaly model given a 3-D density anomaly structure. The second, semi-analytic, approach can be used to solve more complicated problems where the development of a fully analytic Green's function is impossible. This involves using the derivative property of the Fourier transform to reduce the PDE and boundary conditions to algebraic equations that can be solved exactly in the transform domain. A more general model can be constructed by taking the Fourier transform of the source, multiplying by the transform domain solution, and performing the inverse transform numerically. Indeed, the only difference between the two methods is that in the first case, the final model is generated by direct convolution, while in the second case, the convolution theorem is used for model generation. When dealing with spatially complex models, the second approach can sometimes be orders of magnitude more computationally efficient, because of the efficiency of the fast Fourier transform algorithm.

This chapter introduces the minimum amount of Fourier analysis needed to understand the solutions to the PDEs provided in the following chapters. A reader not familiar with Fourier transforms and complex analysis should first study any of the excellent books on the topic. We recommend the first six chapters of the book by Bracewell (1978) for a more complete discussion of the material presented here.

2.2 Definitions of Fourier Transforms

The one-dimensional forward and inverse Fourier transforms are defined as

$$F(k) = \int_{-\infty}^{\infty} f(x) e^{-i2\pi kx} \, dx \quad \text{or} \quad F(k) = \Im\left[f(x)\right] \tag{2.1}$$

$$f(x) = \int_{-\infty}^{\infty} F(k) \, e^{i2\pi kx} \, dk \quad \text{or} \quad f(x) = \Im^{-1}\left[F(k)\right] \tag{2.2}$$

where x is the distance and k is the wavenumber, and where $k = 1/\lambda$ and λ is the wavelength. We also use the shorthand notation introduced by Bracewell (1978). The two-dimensional forward and inverse Fourier transforms are defined as

$$F(\mathbf{k}) = \int_{-\infty}^{\infty} \int_{-\infty}^{\infty} f(\mathbf{x}) \, e^{-i2\pi \mathbf{k} \cdot \mathbf{x}} d^2\mathbf{x} \quad \text{or} \quad F(\mathbf{k}) = \Im_2\left[\mathbf{f}(\mathbf{x})\right] \tag{2.3}$$

$$f(\mathbf{x}) = \int_{-\infty}^{\infty} \int_{-\infty}^{\infty} F(\mathbf{k}) \, e^{i2\pi \mathbf{k} \cdot \mathbf{x}} d^2\mathbf{k} \quad \text{or} \quad f(\mathbf{x}) = \Im_2^{-1}\left[\mathbf{F}(\mathbf{k})\right] \tag{2.4}$$

where $\mathbf{x} = (x, y)$ is the position vector, $\mathbf{k} = (k_x, k_y)$ is the wavenumber vector, and $\mathbf{k} \cdot \mathbf{x} = k_x x + k_y y$. For several of the derivations, we'll also take the Fourier transform in the z-direction (i.e., a 3-D transform) using the following notation

$$F(k_z) = \int_{-\infty}^{\infty} f(z) \, e^{-i2\pi k_z z} \, dz \tag{2.5}$$

$$f(z) = \int_{-\infty}^{\infty} F(k_z) \, e^{i2\pi k_z z} \, dk_z. \tag{2.6}$$

Fourier transformation with respect to time is also sometimes used to form a 4-D transform

$$F(v) = \int_{-\infty}^{\infty} f(t)\, e^{-i2\pi vt}\, \mathrm{dt} \tag{2.7}$$

$$f(t) = \int_{-\infty}^{\infty} F(v)\, e^{i2\pi vt}\, \mathrm{dv}. \tag{2.8}$$

While algebraic manipulation of equations in 4-D is sometimes challenging and error-prone, we'll use computers to help us in two ways. First, we'll use the tools of computer algebra to solve the most challenging algebraic manipulations associated with the 3-D and 4-D problems. This will result in a closed-form solution on the Fourier domain called a *transfer function*. Then we'll use the Fast Fourier Transform (FFT) algorithm to forward transform a complicated source and inverse transform the transfer function times this source, to arrive at the final result.

Note that the wavenumber can be defined to include the 2π. For example, let the wavenumber $s = 2\pi/\lambda$. In this case the forward and inverse transform become

$$F(s) = \int_{-\infty}^{\infty} f(x)\, e^{-isx}\, \mathrm{dx} \tag{2.9}$$

$$f(x) = \frac{1}{2\pi} \int_{-\infty}^{\infty} F(s)\, e^{isx}\, \mathrm{ds}. \tag{2.10}$$

We will use this alternate notation in Chapter 8, and it is commonly used throughout the literature so the reader should be familiar with both notations.

2.3 Fourier Sine and Cosine Transforms

Here we introduce the sine and cosine transforms to illustrate the transforms of odd and even functions. Also, in later chapters, we'll use sine and cosine transforms to match asymmetric and symmetric boundary conditions for particular models.

Any function $f(x)$ can be decomposed into odd $O(x)$ and even $E(x)$ functions such that

$$f(x) = E(x) + O(x) \tag{2.11}$$

where $E(x) = \frac{1}{2}[f(x) + f(-x)]$ and $O(x) = \frac{1}{2}[f(x) - f(-x)]$. Note that the complex exponential function can be written as

$$e^{i\theta} = \cos(\theta) + i\sin(\theta). \tag{2.12}$$

Exercise 2.1 Use equation (2.12) to show that

$$\cos(\theta) = \frac{1}{2}\left(e^{i\theta} + e^{-i\theta}\right) \quad \text{and} \quad \sin(\theta) = \frac{1}{2i}\left(e^{i\theta} - e^{-i\theta}\right).$$

Using this expression (2.12), we can write the forward 1-D transform as the sum of two parts

$$F(k) = \int_{-\infty}^{\infty} f(x)\cos(2\pi kx)\,dx - i\int_{-\infty}^{\infty} f(x)\sin(2\pi kx)\,dx. \tag{2.13}$$

After inserting equation (2.11) into this expression and noting that the integral of an odd function times an even function is zero, we arrive at the expressions for the cosine and sine transforms

$$F(k) = 2\int_{0}^{\infty} E(x)\cos(2\pi kx)\,dx - 2i\int_{0}^{\infty} O(x)\sin(2\pi kx)\,dx. \tag{2.14}$$

Throughout this book, we'll be dealing with real-valued functions. From equation (2.14) it is evident that the cosine transform of a real, even function is also real and even. Also, the sine transform of a real odd function is imaginary and odd. In other words, when a function in the space domain is real valued, its Fourier transform $F(k)$ has a special Hermitian property $F(k) = \overline{F}(-k)$, where the overbar signifies the complex conjugate. Therefore, one can reconstruct the transform of the function with negative wavenumbers from the transform with positive wavenumbers. Later, when we perform numerical examples using real-valued functions such as topography, we can use this Hermitian property to reduce the memory allocation for the Fourier-transformed array by a factor of 2. This is important for large 2-D and 3-D transforms.

2.4 Examples of Fourier Transforms

Throughout the book, we will work with only linear partial differential equations, so all the problems are separable and the order of differentiation and integration is irrelevant. For example, the 2-D Fourier transform is given by:

$$F(k_x, k_y) = \int\limits_{-\infty}^{\infty} \left[\int\limits_{-\infty}^{\infty} f(x,y) e^{-i2\pi k_x x} \, dx \right] e^{-i2\pi k_y y} \, dy$$

$$= \int\limits_{-\infty}^{\infty} \left[\int\limits_{-\infty}^{\infty} f(x,y) e^{-i2\pi k_y y} \, dy \right] e^{-i2\pi k_x x} \, dx. \tag{2.15}$$

Note that this 2-D transform consists of a sequence of 1-D transforms. This property can be extended to 3-D, 4-D, and even N-D; each transform can be performed separately and independently of the transforms in the other dimensions. In the following analysis, we'll only show examples of 1-D transforms, but the extension to higher dimensions is trivial.

Delta Function   By definition the delta function has the following property

$$\int\limits_{-\infty}^{\infty} f(x)\,\delta(x-a)\,dx \equiv f(a). \tag{2.16}$$

Under integration it extracts the value of $f(x)$ at the position $x = a$. One can describe the delta function as having infinite height at zero argument and zero height elsewhere. The area under the delta function is 1. In terms of pure mathematics, the delta function is not a function and only has meaning when integrated against another function. In this book, we use the delta function as a powerful tool provided to us by the mathematicians, so we trust all the mathematical theory behind it. What is the Fourier transform of a delta function? By definition, if one performs a forward transform of a function followed by an inverse transform, or vice versa, one will arrive back with the original function. Let's try this using the delta function. By definition, the inverse transform of a delta function is

$$\int\limits_{-\infty}^{\infty} \delta(k-k_o) e^{i2\pi kx} \, dk = e^{i2\pi k_o x}. \tag{2.17}$$

Next let's take the forward transform of equation (2.17). The left-hand side will be the delta function, because we have performed an inverse and forward transform. The right-hand side is given by

$$\delta(k-k_o) = \int\limits_{-\infty}^{\infty} e^{i2\pi k_o x} e^{-i2\pi kx} \, dx = \int\limits_{-\infty}^{\infty} e^{-i2\pi (k-k_o)x} \, dx. \tag{2.18}$$

This result shows that the Fourier basis functions are orthonormal. If we consider the special case of $k_o = 0$, we see that the inverse Fourier transform of a delta function is $\Im^{-1}[\delta(k)] = 1$. Since Fourier transformation is reciprocal in distance x and wavenumber k, it is also true that $\Im[\delta(x)] = 1$. The delta function and its Fourier transform provide an amazingly powerful tool for solving linear PDEs.

Cosine and Sine Functions   Let's use the delta function tool and the expressions from Exercise 2.1 to calculate the Fourier transform of a cosine function having a single wavenumber $\cos(2\pi k_0 x)$

$$\int_{-\infty}^{\infty} \cos(2\pi k_o x)\, e^{-i2\pi kx}\, \mathrm{d}x = \frac{1}{2} \int_{-\infty}^{\infty} \left(e^{i2\pi k_o x} + e^{-i2\pi k_o x} \right) e^{-i2\pi kx}\, \mathrm{d}x$$

$$= \frac{1}{2} \left[\delta(k - k_o) + \delta(k + k_o) \right]. \tag{2.19}$$

So the Fourier transform of a cosine function is simply two delta functions located at $\pm k_o$.

Exercise 2.2 Show that the Fourier transform of $\sin(2\pi k_0 x)$ is

$$\frac{1}{2i} \left[\delta(k - k_o) - \delta(k + k_o) \right].$$

Gaussian Function   The Gaussian $e^{-\pi x^2}$ function also plays a fundamental role in solutions to several types of PDEs. Its Fourier transform is

$$F(k) = \int_{-\infty}^{\infty} e^{-\pi x^2} e^{-i2\pi kx}\, \mathrm{d}x = \int_{-\infty}^{\infty} e^{-\pi (x^2 + i2kx)}\, \mathrm{d}x. \tag{2.20}$$

Note that $(x + ik)^2 = (x^2 + i2kx) - k^2$. Using this, we can rewrite equation (2.20) as

$$F(k) = e^{-\pi k^2} \int_{-\infty}^{\infty} e^{-\pi (x+ik)^2}\, \mathrm{d}x = e^{-\pi k^2} \int_{-\infty}^{\infty} e^{-\pi (x+ik)^2}\, \mathrm{d}(x + ik) = e^{-\pi k^2} \tag{2.21}$$

where we have used the result that the infinite integral of $e^{-\pi x^2}$ is 1. This is a remarkable and powerful result that the Fourier transform of a Gaussian is simply a Gaussian. See Figure 2.1.

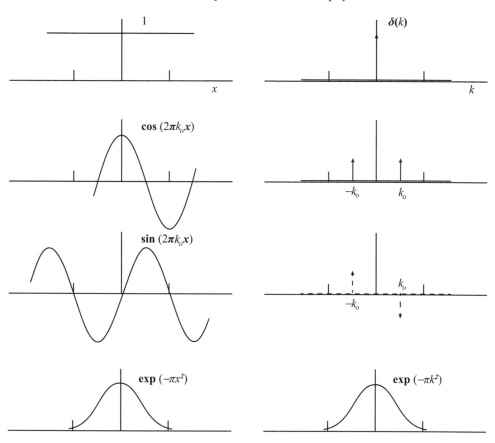

Figure 2.1 Schematic plots of 1-D Fourier transform pairs. The solid lines indicate real-valued functions, while dashed lines indicate imaginary valued functions (modified from Bracewell (1978)).

2.5 Properties of Fourier Transforms

There are several properties of Fourier transforms that can be used as tools for solving PDEs.

Similarity Property   The first property, called the *similarity property*, says that if you scale a function by a factor of a along the x-axis, its Fourier transform will be scaled by a^{-1} along the k-axis and the amplitude will be scaled by $|a|^{-1}$.

> **Exercise 2.3** Use the definition of the Fourier transform equation (2.1) and a change of variable to show the following. Try positive and negative values of a to understand why the absolute value is needed in the amplitude scaling.
>
> $$\Im[f(ax)] = \frac{1}{|a|}F\left(\frac{k}{a}\right) \qquad (2.22)$$

Shift Property   The *shift property* says that the Fourier transform of a function that is shifted by a along the x-axis equals the original Fourier transform scaled by a phase factor. This property is especially useful for numerically shifting a function a non-integer amount of the data spacing along the axis.

> **Exercise 2.4** Use the definition of the Fourier transform and a change of variable to show the following is true.

$$\Im\left[f(x - a)\right] = e^{-i2\pi ka} F(k) \tag{2.23}$$

Derivative Property   The *derivative property* of the Fourier transform is the essential tool used in this book to transform linear PDEs into algebraic equations that are easily solved. It states that the Fourier transform of the derivative of a function is the Fourier transform of the original function scaled by the imaginary wavenumber.

$$\Im\left[\frac{\partial f}{\partial x}\right] = i2\pi k \, F(k) \tag{2.24}$$

To show this is true, we start with the inverse transform of equation (2.24)

$$\frac{\partial f}{\partial x} = \int_{-\infty}^{\infty} i2\pi k \, F(k) e^{i2\pi kx} \, dk. \tag{2.25}$$

Next, take the forward transform of equation (2.25) and rearrange terms

$$\Im\left[\frac{\partial f}{\partial x}\right] = \int_{-\infty}^{\infty}\int_{-\infty}^{\infty} i2\pi k_o \, F(k_o) e^{i2\pi k_o x} \, dk_o e^{-i2\pi kx} \, dx$$

$$= \int_{-\infty}^{\infty} i2\pi k_o \, F(k_o) \left\{ \int_{-\infty}^{\infty} e^{-i2\pi (k-k_o)x} \, dx \right\} dk_o. \tag{2.26}$$

The term in the curly brackets is the delta function $\delta(k - k_o)$ given in equation (2.18). The result is

$$\Im\left[\frac{\partial f}{\partial x}\right] = \int_{-\infty}^{\infty} i2\pi k_o \, F(k_o)\delta(k - k_o) \, dk_o = i2\pi k \, F(k). \tag{2.27}$$

Convolution Theorem   The final property considered here is the *convolution theorem*, which states that the Fourier transform of the convolution of two functions is equal to the product of the Fourier transforms of the original functions.

$$\Im \left[\int_{-\infty}^{\infty} f(u)g(x-u)\,du \right] = F(k)G(k) \tag{2.28}$$

To show this is true, one can perform the Fourier integration on the left side of equation (2.28) and rearrange the order of the integrations

$$\Im \left[\int_{-\infty}^{\infty} f(u)g(x-u)\,du \right] = \int_{-\infty}^{\infty} \left[\int_{-\infty}^{\infty} f(u)g(x-u)\,du \right] e^{-i2\pi kx}\,dx$$

$$= \int_{-\infty}^{\infty} f(u) \left\{ \int_{-\infty}^{\infty} g(x-u)e^{-i2\pi kx}\,dx \right\} du. \tag{2.29}$$

Next, use the shift property of the Fourier transform to note that the function in the curly brackets on the right side of equation (2.29) is $e^{-i2\pi ku}G(k)$. The result becomes

$$\Im \left[\int_{-\infty}^{\infty} f(u)g(x-u)\,du \right] = G(k) \int_{-\infty}^{\infty} f(u)\,e^{-i2\pi ku}\,du = F(k)G(k). \tag{2.30}$$

Note that these four properties are equally valid in two dimensions or even N dimensions. The properties also apply to discrete data. See Chapter 18 in Bracewell (1978).

Cauchy Residue Theorem   The *Cauchy residue theorem* is an additional tool that we will use many times in the book to perform inverse transforms for cases where the function is analytic and has poles in the complex plane. Let $f(z)$ be an analytic function in the complex plane $z = x + iy$.

An analytic function has the special property that a path integral of the function about any closed loop in the complex plane is zero

$$\oint f(z)\,dz = 0. \tag{2.31}$$

As an example, the gravitational potential associated with the topography of the Earth represents an analytic function. A cyclist riding along any closed path will gain and lose potential energy along the path, but no matter how the path is traversed, they will have the same potential at the end of the circuit as they started with. Next, suppose the same integration is performed with a pole (zero) in the denominator at a complex point z_o. The Cauchy residue theorem states

$$\oint \frac{f(z)}{z - z_o}\, dz = i2\pi f(z_o).$$

(2.32)

The function in the numerator could be very complicated, but as long as it is analytic, the path integral can be evaluated.

Exercise 2.5 Without using the Cauchy residue theorem, show that the following is true.

$$\oint \frac{1}{z}\, dz = i2\pi$$

(2.33)

2.6 Solving a Linear PDE Using Fourier Methods and the Cauchy Residue Theorem

In the following chapters, we'll derive the Green's function and/or its Fourier transform, starting from a PDE and boundary conditions. The approach will follow the same format, so here we have selected a simple example to illustrate the general method.

Heat Flow for a Line Source of Heat at Depth

Consider a line source of heat at a depth of $-a$ buried in a conductive half space, as shown in Figure 2.2,

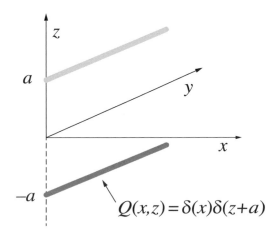

Figure 2.2 Line source of heat at $z = -a$. The surface boundary condition can be met by placing an equal but opposite line sink of heat at $z = a$.

Calculate the temperature in the half space. The differential equation and three boundary conditions are

$$\frac{\partial^2 T}{\partial x^2} + \frac{\partial^2 T}{\partial z^2} = \delta(x)\delta(z+a)$$

$$T(x,0) = 0$$

$$\lim_{|x|\to\infty} T(x,z) = 0 \qquad\qquad (2.34)$$

$$\lim_{z\to\infty} T(x,z) = 0.$$

The first step in the solution is to take the 2-D Fourier transform of the PDE. Each derivative on the left hand side is replaced with an $i2\pi k$ according to the derivative property of the Fourier transform, and the 2-D transform of the right hand side is done using the definition of the Delta function. The result is

$$-4\pi^2\left(k_x^2 + k_z^2\right)T(k_x, k_z) = e^{i2\pi k_z a}$$

$$T(k_x, k_z) = \frac{e^{i2\pi k_z a}}{-4\pi^2\left(k_x^2 + k_z^2\right)}. \qquad\qquad (2.35)$$

Now that we have solved the algebraic problem, we'll start by taking the inverse Fourier transform in the z-direction.

$$T(k_x, z) = \frac{-1}{4\pi^2} \int_{-\infty}^{\infty} \frac{e^{i2\pi k_z(z+a)}}{\left(k_z^2 + k_x^2\right)}\, dk_z \qquad\qquad (2.36)$$

The denominator of equation (2.36) can be factored as $(k_z + ik_x)(k_z - ik_x)$, which represent two poles in the complex plane (Figure 2.3).

$$T(k_x, z) = \frac{-1}{4\pi^2} \int_{-\infty}^{\infty} \frac{e^{i2\pi k_z(z+a)}}{(k_z + ik_x)(k_z - ik_x)}\, dk_z \qquad\qquad (2.37)$$

To perform this integration, we'll integrate around one of these poles using the Cauchy residue theorem. First consider the case $k_x > 0$, $z > -a$. We would like to integrate along the k_z-axis from $-\infty$ to ∞, as shown in Figure 2.3. If we close the integration path in the upper hemisphere of the complex plane, the numerator will become vanishingly small, because it is a decaying exponential function. Therefore, the integration along the k_z-axis will be equivalent to the full integration around the pole $k_z = ik_x$. Using equation (2.32), the result is

$$T(k_x, z) = \frac{-i2\pi}{4\pi^2} \frac{e^{i2\pi ik_x(z+a)}}{2ik_x} = \frac{-e^{-2\pi k_x(z+a)}}{4\pi k_x}. \qquad\qquad (2.38)$$

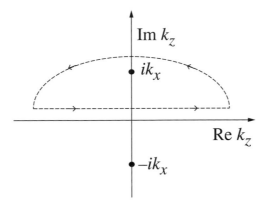

Figure 2.3 Closed integration path about a pole in the complex plane.

Note that this is a decaying exponential function of z that satisfies the last boundary condition in equation (2.34). Next, consider the case $k_x < 0$, $z > -a$. This time, we close the integration path in the lower hemisphere to also achieve a decaying exponential function. Integration about the pole in the clockwise direction reverses the sign in the Cauchy residue theorem, so the result is

$$T(k_x, z) = \frac{-i2\pi}{4\pi^2} \frac{e^{-i2\pi i k_x(z+a)}}{-2i k_x} = \frac{e^{2\pi k_x(z+a)}}{4\pi k_x}. \tag{2.39}$$

Equations (2.38) and (2.39) can be combined by using the absolute value of k_x.

$$T(k_x, z) = \frac{-e^{-2\pi |k_x|(z+a)}}{4\pi |k_x|} \tag{2.40}$$

The next step is to take the inverse transform with respect to k_x.

$$T(x, z) = \frac{-1}{4\pi} \int_{-\infty}^{\infty} \frac{e^{-2\pi |k_x|(z+a)}}{|k_x|} e^{i2\pi k_x x} \, dk_x \tag{2.41}$$

One way to perform this final integration is to use the derivative property of the Fourier transform to find the solution for $\partial T/\partial z$ and then integrate over z to get the desired result.

$$\frac{\partial T(x, z)}{\partial z} = \frac{1}{2} \int_{-\infty}^{\infty} e^{-2\pi |k_x|(z+a)} e^{i2\pi k_x x} \, dk_x \tag{2.42}$$

One can look up this definite integral

$$\frac{\partial T(x, z)}{\partial z} = \frac{-(z+a)}{4\pi [x^2 + (z+a)^2]}, \quad z > -a. \tag{2.43}$$

After integrating over z, we find

$$T(x,z) = \frac{-1}{4\pi} \ln [x^2 + (z+a)^2]^{1/2}. \tag{2.44}$$

Finally, we have not yet met the surface boundary condition $T(x,0) = 0$. This can be achieved by placing a line heat sink at $z = a$. The sum of the source and sink satisfies the differential equation and the four boundary conditions.

$$T(x,z) = \frac{-1}{4\pi} \left\{ \ln [x^2 + (z+a)^2]^{1/2} - \ln [x^2 + (z-a)^2]^{1/2} \right\} \tag{2.45}$$

2.7 Fourier Series

Many geophysical problems are concerned with a small area on the surface of the Earth having a width of W and length of L, as shown in Figure 2.4.

The coefficients of the two-dimensional Fourier series are computed by the following integration

$$F_n^m = \frac{1}{LW} \int_0^L \int_0^W f(x,y) \exp\left[-i2\pi \left(\frac{m}{L}x + \frac{n}{W}y\right)\right] dy \, dx. \tag{2.46}$$

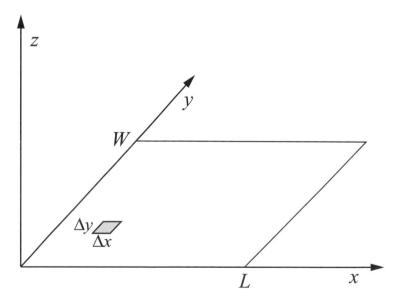

Figure 2.4 Cartesian coordinate system used throughout the book with z positive up. The $z = 0$ plane is the surface of the Earth. Fourier transforms deal with infinite domains while the Fourier series has finite domains. For our numerical examples we will select an area of length L and width W consisting of uniform cells of size Δx and Δy. This can be represented as a 2-D array of numbers with $J = L/\Delta x$ columns and $I = W/\Delta y$ rows.

The function is reconstructed by the following summations over the Fourier coefficients

$$f(x, y) = \sum_{n=-\infty}^{\infty} \sum_{m=-\infty}^{\infty} F_n^m \exp\left[i2\pi \left(\frac{m}{L}x + \frac{n}{W}y\right)\right]. \tag{2.47}$$

The finite size of the area leads to a discrete set of wavenumbers $k_x = m/L, k_y = n/W$, and a discrete set of Fourier coefficients F_n^m. In addition to the finite size of the area, geophysical data commonly have a characteristic sampling interval Δx and Δy. Note that $I = L/\Delta x$ is the number of points in the x-direction and $J = W/\Delta y$ is the number of points in the y-direction. The Nyquist wavenumbers are $k_x = 1/(2\Delta x)$ and $k_x = 1/(2\Delta x)$, so there is a finite set of Fourier coefficients $-I/2 < m < I/2$ and $-J/2 < n < J/2$. Recall the trapezoidal rule of integration

$$\int_0^L f(x)\, dx \cong \sum_{i=0}^{I-1} f(x_i) \Delta x \quad \text{where } x_1 = i\Delta x$$

$$\int_0^L f(x)\, dx \cong \frac{L}{I} \sum_{i=0}^{I-1} f(x_i). \tag{2.48}$$

The discrete forward and inverse Fourier transforms are

$$F_n^m = \frac{1}{IJ} \sum_{i=0}^{I-1} \sum_{j=0}^{J-1} f_i^j \exp\left[-i2\pi \left(\frac{m}{I}i + \frac{n}{J}j\right)\right] \tag{2.49}$$

$$f_i^j = \sum_{n=-I/2}^{I/2-1} \sum_{m=-J/2}^{J/2-1} F_n^m \exp\left[i2\pi \left(\frac{i}{I}m + \frac{j}{J}n\right)\right]. \tag{2.50}$$

The summations for the forward and inverse discrete Fourier transforms are similar, so one can use the same computer code for both transforms. Sorry for the dual use of the letter 'i.' The italic 'i' in front of the 2π is $\sqrt{-1}$, whereas the non-italic 'i's are integers.

2.8 Exercises

Exercise 2.6 What is the Fourier transform of the following function? Show your work and simplify the result.

$$\Pi(x) = \begin{cases} 1 & |x| < 1/2 \\ 1/2 & |x| = 1/2 \\ 0 & otherwise \end{cases} \tag{2.51}$$

Exercise 2.7 Use the convolution theorem to calculate the Fourier transform of the following. Show your work.

$$\Lambda(x) = \begin{cases} 1 - |x| & |x| \leq 1 \\ 0 & |x| > 1 \end{cases} \tag{2.52}$$

Note $\Lambda = \Pi * \Pi$.

Exercise 2.8 Perform the following path integral on $|z| = 2$.

$$\oint \frac{z+2}{(z^2+1)} \, dz \tag{2.53}$$

Exercise 2.9

```
%
% 1)   Write a program to generate a cosine function
%        using 2048 points. Generate exactly 32, or 64 cycles
%        of the function. Plot the results and add labels.
%
figure(1)
clf
nx=2048;
kc=64/nx;
x=0:nx-1;
%
%   generate the function
%
y=cos(2*pi*x*kc);
%
figure(1)
plot(x,y);
xlabel('x')
ylabel('cos(x)')
pause
%
% 2) Take the Fourier transform of the function that you made in
%        problem 1. Use fftshift to shift the zero frequency to the center
%        of the spectrum. Generate wavenumbers for the horizontal axis.
%        Take the inverse FFT. Do you get what you started with? (don't
%        forget to undo the fftshift.)
%
figure(2)
subplot(5,1,1),plot(x,y);
xlabel('x')
ylabel('cos(x)')
%
%   generate the wavenumbers
%
k=-nx/2:nx/2-1;
%
cy=fftshift(fft(y));
subplot(5,1,2),plot(k,real(cy));
xlabel('k')
subplot(5,1,3),plot(k,imag(cy));
%
%   do the inverse FFT
%
```

```
yo=ifft(fftshift(cy));
subplot(5,1,4),plot(x,real(yo));
xlabel('x')
ylabel('cos(x)')
subplot(5,1,5),plot(x,real(y-yo));
xlabel('x')
ylabel('difference')
pause
%
% 3) Do problem 2 over, using a sine function instead of a cosine
%    function.
%
% 4) Show that the Fourier transform of a Gaussian function is a
%    Gaussian function. Plot the difference between the fft result and
%    the exact function. When you do this problem, it is best to make
%    the Gaussian function an even function of x just prior to
%    computing the fft(). If you do this then the transformed Gaussian
%    will be real and even. Also you will need to scale the transform
%    by the point spacing dx = L/nx.
%
clear
figure(3)
nx=2048;
L=20;
dx=L/nx;
a=1.;
x=a*(-nx/2:nx/2-1)*dx;
g=exp(-pi*x.*x);
subplot(4,1,1),plot(x,g);
axis([-4,4,-.5,1.1])
xlabel('x')
ylabel('Gaussian')
%
%  generate the wavenumbers
%
k=(-nx/2:nx/2-1)/L;
%
cg=fftshift(fft(fftshift(g)))*dx;
%
% 5) Use this Gaussian example to demonstrate the stretch property of
%    Fourier transform. The results should be compared in the
%    wavenumber domain.
%
%
% 6) Use this Gaussian function to illustrate the shift property of
%    the Fourier transform. The results should be displayed as a
%    shifted Gaussian in the space domain.
%
%
% 7) Use the Gaussian function to demonstrate the derivative property
%    of the Fourier transform. The analytic derivative of the Gaussian
%    should be compared with the Fourier derivative in the space
%    domain.
%
```

3

Plate Kinematics

3.1 Plate Motions on a Flat Earth

Plate tectonic theory describes the motions of rigid plates on a spherical Earth. However, when considering the relative motions very close to the plate boundary or at a triple junction, it is appropriate to use a flat-Earth approximation. We'll begin with the flat-Earth case and then move on to the spherical case (Fowler, 1990). Consider the two plates A and B, which have a subduction zone boundary between them, such as the Nazca and South American plates. In this analysis, all plate motions are relative, so one can either consider plate B as fixed or plate A as fixed, and draw the relative vector velocity between them—as shown in Figure 3.1.

3.2 Triple Junction

A triple junction is the intersection of three plate boundaries. The most common types of triple junctions are ridge-ridge-ridge (R-R-R), ridge-fault-fault (R-F-F), and ridge-trench-trench (R-T-T); see Figure 3.2.

Each type of plate boundary has rules about relative velocities:

Ridge relative velocity must be divergent and is usually perpendicular to the ridge.
Transform Fault relative velocity must be parallel to the fault.
Trench relative velocity must be convergent, but no direction is preferred.

All triple junctions must satisfy a velocity condition such that the vector sum around the plate circuit is zero.

$$\mathbf{V}_{BA} + \mathbf{V}_{CB} + \mathbf{V}_{AC} = 0 \tag{3.1}$$

In most cases we can map the geometry of the spreading ridges, transform faults, and trenches, but cannot always measure the relative velocities. The triple junction closure equation (3.1) can be used to solve for spreading velocities given the triple junction geometry, the rules, and at least one relative plate velocity.

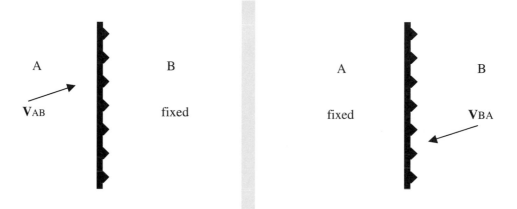

$\mathbf{V}_{AB}$ velocity vector of plate A relative to plate B.

$\mathbf{V}_{BA}$ velocity vector of plate B relative to plate A.

$$\mathbf{V}_{AB} = -\mathbf{V}_{BA}$$
$$\mathbf{V}_{BA} = V_x i + V_y j$$

Figure 3.1

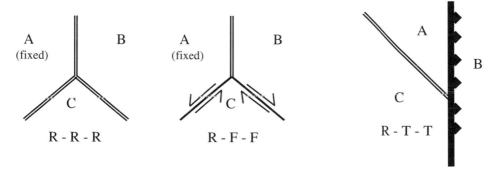

Figure 3.2 Three of the most common triple junctions: R-ridge, F-fault, and T-trench.

Example 3.1 Given the geometry of the Indian Ocean Triple Junction in Figure 3.3 and one spreading rate $|\mathbf{V}_{CB}| = 50$ mm/yr, calculate the other two spreading rates.

The first step is to construct a diagram of the sum of the relative plate velocities as shown in Figure 3.4. In this case we used the rule that the relative spreading direction between two plates is described by the orientation of the significant-offset transform faults closest to the triple junction (dashed lines in Figure 3.3). We then use the law of sines to solve for the lengths of the vectors on the other two sides of the triangle.

$$\frac{|\mathbf{V}_{BA}|}{\sin 45} = \frac{|\mathbf{V}_{AC}|}{\sin 17} = \frac{|\mathbf{V}_{CB}|}{\sin 118} \tag{3.2}$$

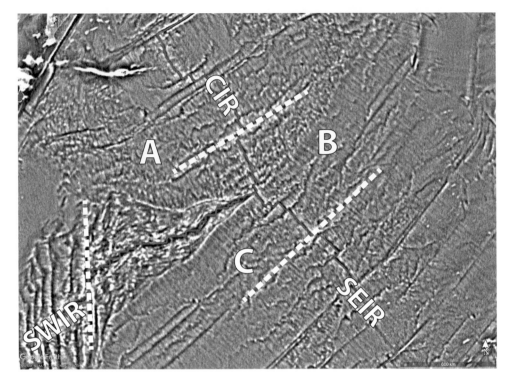

Figure 3.3 Indian Ocean Triple Junction: R-R-R. The three spreading ridges are the Central Indian Ridge (CIR), the Southeast Indian Ridge (SEIR), and the Southwest Indian Ridge (SWIR). The greyscale image is the vertical gravity gradient discussed in later chapters.

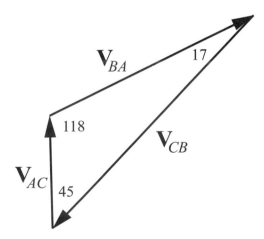

Figure 3.4 Sum of interior angles $= 180°$

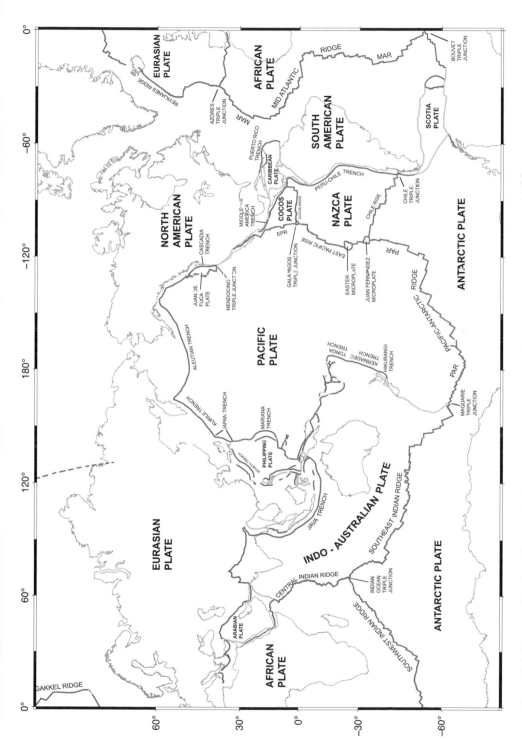

Figure 3.5 Names of major plates and plate boundaries modified from Fowler (1990). Ridges are red, transforms are green, and trenches are blue. (For a color version of this figure, please see the color plate section.)

We find $|\mathbf{V}_{BA}| = 40.0$ mm/yr and $|\mathbf{V}_{AC}| = 16.6$ mm/yr. The map in Figure 3.5 shows the other triple junctions. As an exercise, use a bathymetric map (e.g., Google Earth and this KMZ-file `topex.ucsd.edu/pub/srtm30_plus/SRTM30_PLUS.kmz`) to determine the geometry of another triple junction, and then use Table 3.1 in the next section to calculate the spreading rate at one of the ridges. The next section develops the mathematics for calculation of plate motions on a spherical Earth.

3.3 Plate Motions on a Sphere

These notes are largely based on three publications (Minster and Jordan, 1978; DeMets et al., 1990, 2010).

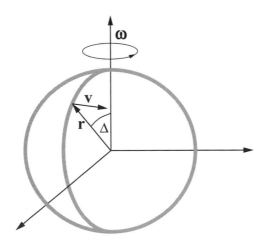

Given:

ω angular velocity vector $\left(\frac{\text{rad}}{\text{s}}\right)$

$\mathbf{r}$ position on Earth (m)

Calculate:

$\mathbf{v}$ velocity vector at position $\mathbf{r}$ $\left(\frac{\text{m}}{\text{s}}\right)$

Of course, the velocity of the plate must be tangent to the surface of the Earth, so the velocity is the cross product of the position vector and the angular velocity vector.

$$\mathbf{v} = \omega \times \mathbf{r} \qquad\qquad (3.3)$$

or

$$\mathbf{v} = \hat{\imath}(\omega_y z - \omega_z y) - \hat{\jmath}(\omega_x z - \omega_z x) + \hat{k}(\omega_x y - \omega_y x) \qquad\qquad (3.4)$$

where $\hat{\imath}$, $\hat{\jmath}$, and $\hat{k}$ are unit vectors. The magnitude of the velocity is given by

$$|\mathbf{v}| = |\omega||\mathbf{r}|\sin(\Delta) \qquad\qquad (3.5)$$

Table 3.1. *Best-fitting angular velocities describe counter-clockwise rotation of the first plate relative to the second. From DeMets et al. (2010).*

Plate pair	Latitude (deg.)	Longitude (deg.)	ω (deg. Myr^{-1})
AU-AN	11.3	41.8	0.633
CP-AN	17.2	32.8	0.580
LW-AN	-1.2	-33.6	0.133
NB-AN	-6.2	-34.3	0.158
NZ-AN	33.1	-96.3	0.477
PA-AN	-65.1	99.8	0.870
SM-AN	11.2	-56.7	0.140
EU-NA	61.8	139.6	0.210
NB-NA	79.2	40.2	0.233
AR-NB	30.9	23.6	0.403
CO-NZ	1.6	-143.5	0.636
CO-PA	37.4	-109.4	2.005
JF-PA	-0.6	37.8	0.625
NZ-PA	52.7	-88.6	1.326
RI-PA	25.7	-104.8	4.966
NB-SA	60.9	-39.0	0.295
SW-SC	-32.0	-32.2	1.316
AR-SM	22.7	26.5	0.429
CP-SM	16.9	45.8	0.570
IN-SM	22.7	30.6	0.408
AN-SR	85.7	-139.3	0.317
NB-SR	70.6	-60.9	0.346

where Δ is the angle between the position vector and the angular velocity vector. It is given by the following formula:

$$\cos(\Delta) = \frac{\boldsymbol{\omega} \cdot \mathbf{r}}{|\boldsymbol{\omega}||\mathbf{r}|} \tag{3.6}$$

The formulas above assume that the angular velocity vector and the position vector are provided in Cartesian coordinates. However, usually they are specified in terms of latitude and longitude. Thus, one must transform both vectors. The usual case is to calculate the relative velocity between two plates somewhere along their common boundary. Table 3.1 lists the pole position and rates of rotation for relative motion between plate pairs shown in Figure 3.6. The Cartesian position of a point along the plate boundary is

$$\begin{aligned} x &= a \cos \theta \cos \phi \\ y &= a \cos \theta \sin \phi \\ z &= a \sin \theta \end{aligned} \tag{3.7}$$

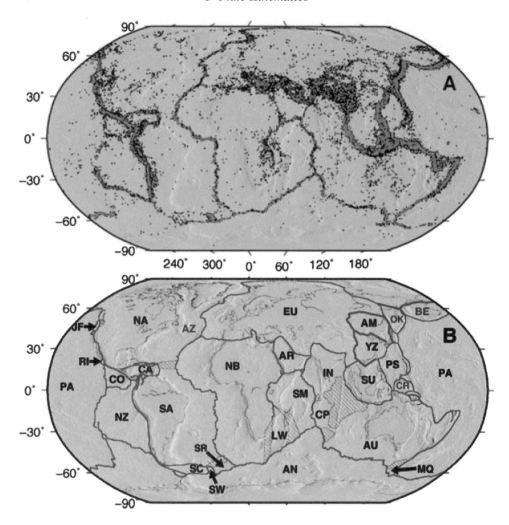

Figure 3.6 (A) Epicenters for earthquakes with magnitudes equal to or larger than 3.5 (black) and 5.5 (dark gray), and depths shallower than 40 km, for the period 1967–2007. Hypocentral information is from US Geological Survey National Earthquake Information Center files. (B) Plate boundaries and geometries employed for MORVEL. Plate name abbreviations are as follows: AM, Amur; AN, Antarctic; AR, Arabia; AU, Australia; AZ, Azores; BE, Bering; CA, Caribbean; CO, Cocos; CP, Capricorn; CR, Caroline; EU, Eurasia; IN, India; JF, Juan de Fuca; LW, Lwandle; MQ, Macquarie; NA, North America; NB, Nubia; NZ, Nazca; OK, Okhotsk; PA, Pacific; PS, Philippine Sea; RI, Rivera; SA, South America; SC, Scotia; SM, Somalia; SU, Sundaland; SW, Sandwich; YZ, Yangtze. Light gray labels indicate plates not included in MORVEL. Patterned Dark gray areas show diffuse plate boundaries. Figure from DeMets et al. (2010).

where θ is latitude, ϕ is longitude, and a is the mean Earth radius. It is helpful to memorize the conversion from latitude-longitude to the Cartesian coordinate system, where the x-axis runs from the center of the Earth to a point at $0°$ latitude and $0°$ longitude (i.e., the Greenwich meridian), the y-axis runs through a point at $0°$ latitude and $90°$ east longitude, and the z-axis runs along the spin-axis to the North Pole.

Similarly, the pole positions must be converted from geographic coordinates (θ_p, ϕ_p) into the Cartesian system:

$$\omega_x = |\boldsymbol{\omega}| \cos \theta_p \cos \phi_p$$
$$\omega_y = |\boldsymbol{\omega}| \cos \theta_p \sin \phi_p \tag{3.8}$$
$$\omega_z = |\boldsymbol{\omega}| \sin \theta_p$$

where $|\boldsymbol{\omega}|$ is the magnitude of the rotation vector provided in Table 3.1. There are two ways to compute the magnitude of the velocity. One could compute the cross product of the rotation vector and the position vector (equation (3.3)). Then the magnitude of the velocity is

$$|\mathbf{v}| = \left(v_x^2 + v_y^2 + v_z^2 \right)^{1/2}. \tag{3.9}$$

One could also calculate the angle Δ between the position vector and the angular velocity vector using equation (3.6) and then use that value in equation (3.5) to calculate the magnitude of the velocity. Indeed, both Fowler (1990) and *Geodynamics* (Turcotte and Schubert, 2014) use this second approach. However, they use the rather cumbersome spherical trigonometry to calculate the angle Δ. We prefer to use equation (3.5), after converting everything to Cartesian coordinates.

3.4 Velocity Azimuth

We know that the velocity vector is tangent to the sphere. Given the Cartesian velocity components from equation (3.4), we would like to compute the latitude v_θ and longitude v_ϕ components of velocity. Begin by taking the derivative of equation (3.7) with respect to time:

$$v_x = a(- \cos \phi \sin \theta \ v_\theta - \cos \theta \sin \phi \ v_\phi)$$
$$v_y = a(- \sin \phi \sin \theta \ v_\theta + \cos \theta \cos \phi \ v_\phi) \tag{3.10}$$
$$v_z = a(\cos \theta \ v_\theta)$$

From the last equation in (3.10), we can solve for the latitude velocity component.

$$v_\theta = \frac{v_z}{a \, \cos \theta} \tag{3.11}$$

Now plug v_θ into either the v_x or v_y equation, and solve for v_ϕ.

$$v_\phi = \frac{v_y + v_z \sin \phi \tan \theta}{a \cos \theta \cos \phi} \tag{3.12}$$

If this equation turns out to be singular, then use the v_x equation:

$$v_\phi = -\frac{v_x + v_z \cos \phi \tan \theta}{a \cos \theta \sin \phi} \tag{3.13}$$

3.5 Recipe for Computing Velocity Magnitude

In summary, to calculate the magnitude of the velocity:

1. Transform lat and lon into the unit vector $\mathbf{x} = (x, y, z)$ using equation (3.7).
2. Transform pole lat and lon into the unit vector $\mathbf{x}_p = (x_p, y_p, z_p)$ using equation (3.8).
3. $\cos \Delta = \mathbf{x} \cdot \mathbf{x}_p$
4. $v = \omega a \sin \Delta$

Example 3.2 Given the rotation pole between the Pacific and Nazca plates, calculate the spreading rate at $-20°\,113.5°\mathrm{W}$.

Pole			Point	
52.7	−88.6	1.326×10^{-6} deg/yr	20.0°S	113.5°W
52.7	271.4	2.314×10^{-8} rad/yr	−20.0	246.5
$x_p = 0.0148$			$x = -0.375$	
$y_p = -0.606$			$y = -0.862$	
$z_p = 0.795$			$z = -0.342$	

$$\cos \Delta = \mathbf{x} \cdot \mathbf{x}_p = (-0.0056 + .522 - .272) = .244$$
$$\Delta = 75.8 \quad v = \omega\, a \sin \Delta = 142.9 \text{ mm/yr}$$

3.6 Triple Junctions on a Sphere

Triple junction closure on a sphere is similar to triple junction closure on a flat-Earth, except that the sum of the rotation vectors must be zero:

$$\omega_{BA} + \omega_{CB} + \omega_{AC} = 0 \tag{3.14}$$

Example 3.3 *Galapagos Triple Junction* Given the rotation vectors of the Cocos plate relative to the Pacific plate and the Pacific plate relative to the Nazca plate, calculate the spreading rate at $2°\mathrm{N}, 260°\mathrm{E}$.

$$\omega_{CP} + \omega_{NC} + \omega_{PN} = 0$$

$$\omega_{NC} = -\omega_{CP} - \omega_{PN}$$

$$\mathbf{v}_{NC} = \omega_{NC} \times \mathbf{r}(\theta, \phi)$$

$|\mathbf{v}|$ is the magnitude of the spreading rate

3.7 Hot Spots and Absolute Plate Motions

So far, we have only considered relative plate motions, because there was no way to tie the positions of the plates to the mantle. One method of making this connection and thus determining absolute plate motions is to assume that "hot spots" remain fixed with respect to the lower mantle.

A hot spot is an area of concentrated volcanic activity. There is a subset of hot spots that have the following characteristics:

1. They produce linear volcanic chains in the interiors of the plates.
2. The youngest volcanoes occur at one end of the volcanic chain, and there is a linear increase in age away from that end.
3. The chemistry of the erupted lavas is significantly different from lava erupted at mid-ocean ridges or island arcs.
4. Some hot spots are surrounded by a broad topographic swell about 1000 m above the surrounding ocean basin.

These features are consistent with a model where the plates are moving over a relatively fixed mantle plume. After identifying the linear volcanic chains associated with the mantle plumes, it has been shown that the relative motions among hot spots is about 10 times less than the relative plate motions.

3.8 Exercises

Exercise 3.1 Calculate the spreading rate at a point on the northern Mid-Atlantic Ridge (latitude 30, longitude 319).

Exercise 3.2 Calculate the slip rate along the San Andreas Fault in San Francisco (latitude 38, longitude −122.7). You will need to use tables from DeMets et al. (2010).

Exercise 3.3 Where is the fastest seafloor spreading ridge on the Earth? Use Table 3.1 to calculate the spreading rate at that location.

Exercise 3.4 The vector sum of relative plate velocities around a triple junction is zero:

$$\mathbf{v}_{BA} + \mathbf{v}_{CB} + \mathbf{v}_{AC} = 0 \tag{3.15}$$

Show that the following is also true at a position $\mathbf{r}_o$:

$$\omega_{BA} + \omega_{CB} + \omega_{AC} = 0 \tag{3.16}$$

where the ω's are the relative rotation poles on a sphere and $\mathbf{r}_o$ is not parallel to any of the ω's.

Exercise 3.5 Use the Google Earth overlays of vertical gravity gradient and earthquake epicenters to sketch the geometry of the ridges and transform faults around the Galapagos triple junction. Given the spreading rate across the southern segment of the East Pacific Rise (EPR) of 120 mm/yr, calculate the spreading rates on the northern segment of the EPR and the Cocos ridge.

4

Marine Magnetic Anomalies

4.1 Introduction

This chapter develops the equations needed to compute the scalar magnetic field that is recorded by a magnetometer towed behind a ship, given a magnetic timescale, a spreading rate, and a skewness (e.g., Schouten (1971); Schouten and McCamy (1972); Gee and Kent (2007)). A number of assumptions are made to simplify the mathematics. The intent is to first review the origin of natural remnant magnetism (NRM), to illustrate that the magnetized layer is thin compared with its horizontal dimension. Then the relevant differential equations are developed and solved under the ideal case of seafloor spreading at the north magnetic pole. This development highlights the Fourier approach to the solution to linear partial differential equations. The same approach will be used to develop the Green's functions for heat flow, flexure, gravity, and elastic dislocation. For a more general development of the geomagnetic solution, see the paper by Parker (1973).

4.2 Crustal Magnetization at a Spreading Ridge

As magma is extruded at the ridge axis, its temperature falls below the Curie point, and the uppermost part of the crust becomes magnetized in the direction of the ambient magnetic field. Figure 4.1, from Kent et al. (1993), illustrates the current model of crustal generation. Partial melt that forms by pressure-release in the uppermost mantle ($\sim$40 km depth) percolates to a depth of about 2000 m beneath the ridge, where it accumulates to form a thin magma lens. Beneath the lens a mush-zone develops into a 3500 m thick gabbro layer, by some complicated ductile flow. Above the lens, sheeted dikes ($\sim$1400 m thick) are injected into the widening crack at the ridge axis. Part of this volcanism is extruded into the seafloor

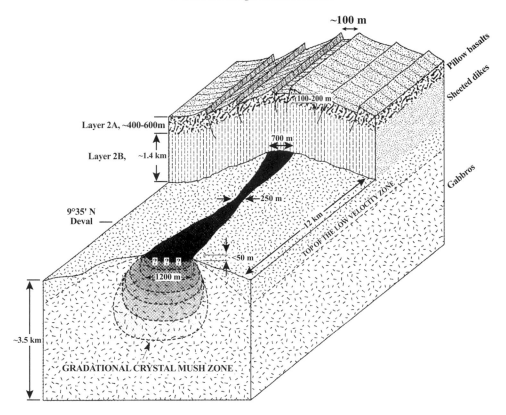

Figure 4.1 Model of crustal structure derived from reflection and refraction seismology (Kent et al., 1993).

as pillow basalts. The pillow basalts and sheeted dikes cool rapidly below the Curie temperature as cool seawater percolates to a depth of at least 2000 m. This process forms the basic crustal layers seen by reflection and refraction seismology methods.

The highest magnetization occurs in the extrusives, forming seismic layer 2A (Figure 4.2, Table 4.1), although the sheeted dikes and gabbro layers provide some contribution to the magnetic anomaly measured on the ocean surface. Note that the reversals recorded in the gabbro layer do not have sharp vertical boundaries (Figure 4.3). The tilting reflects the time delay when the temperature of the gabbro falls below the Curie point. The sea-surface magnetic-anomaly model shown in Figure 4.3 (Gee and Kent, 1994) includes the thickness and precise geometry of the magnetization of all three layers. For the calculation below, we assume all of the magnetic field comes from the thin extrusive layer.

Table 4.1.

Layers	Thickness Seismic Velocity	Description	Thermoremnant Magnetism (TRM)
layer 1	variable <2.5 km/s	sediment	N/A
layer 2A	400–600 m 2.2–5.5 km/s	extrusive, pillow basalts	5–10 A m^{-1}
layer 2B	1400 m 5.5–6.5 km/s	intrusive, sheeted dikes	~1 A m^{-1}
layer 3	3500 m 6.8–7.6 km/s	intrusive, gabbro	~1 A m^{-1}

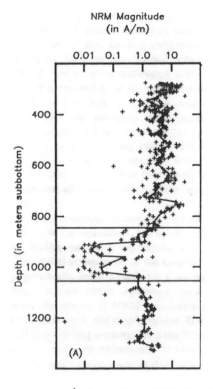

Figure 4.2 NRM values (in Am^{-1}) from Hole 504B. Depths are measured from the seabottom and include 274.5 m of sediment. The horizontal lines separate the upper units, the transition zone, and the dike complex. (From Smith and Banerjee (1986).)

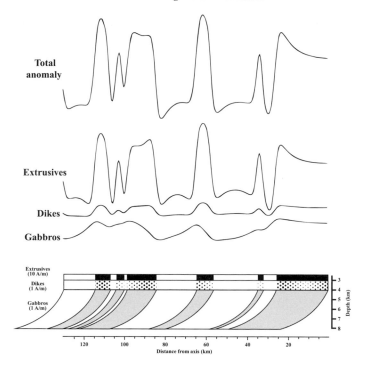

Figure 4.3 Magnetic anomalies generated by a realistic model of crustal magnetization at a half spreading rate of 32 mm/yr. The primary magnetization signature
comes from the thin layer of extrusives. Dipping magnetization in the Gabbros
reflects the position of the Curie isotherm at depth away from the ridge axis.
(Jeff Gee, personal communication.)

The other assumptions are:

1. The ridge axis is 2-D, so there are no along-strike variations in magnetization.
2. The magnetization contrast between alternately magnetized blocks is sharp
   relative to the mean ocean depth.
3. The spreading rate is uniform with time. Before going into the calculation, we
   briefly review the magnetic field generated by a uniformly magnetized block.

4.3 Uniformly Magnetized Block

$\mathbf{M}$ magnetization vector (A m^{-1})

$\mathbf{\Delta}B$ magnetic anomaly vector (T)

$\boldsymbol{\mu}_o$ magnetic permeability ($4\pi \times 10^{-7}$ T A^{-1}m)

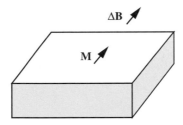

A magnetized rock contains minerals of magnetite and hematite that can be preferentially aligned in some direction. For a body with a uniform magnetization direction, the magnetic anomaly vector will be parallel to that direction. The amplitude of the external magnetic field will have some complicated form:

$$\Delta \mathbf{B}(\mathbf{r}) = \mu_o \mathbf{M} f(\mathbf{r}) \tag{4.1}$$

where $f(\mathbf{r})$ is a function of position that depends on geometry. The total magne tization of a rock has two components: thermoremnant magnetism (TRM) $\mathbf{M}_{TRM}$, and magnetization that is induced by the present-day dipole field $\mathbf{M}_I$:

$$\mathbf{M} = \mathbf{M}_{TRM} + \mathbf{M}_I \qquad \mathbf{M}_I = \chi \mathbf{H} \tag{4.2}$$

where χ is the magnetic susceptibility and H is the applied dipole field of the Earth. The Koenigberger ratio Q is the ratio of the remnant field to the induced field. This value should be much greater than 1 to be able to detect the crustal anomaly. Like the magnetization, the value of Q is between 5 and 10 in Layer 2A, but falls to about 1 deeper in the crust.

4.4 Anomalies in the Earth's Magnetic Field

When a magnetometer is towed behind a ship, one measures the total magnetic field $\mathbf{B}$, and must subtract out the reference Earth magnetic B_e field to establish the magnetic anomaly ΔB:

$$\mathbf{B} = \mathbf{B}_e + \Delta \mathbf{B} \tag{4.3}$$

Most marine magnetometers measure the scalar magnetic field. This is an easier measurement, because the orientation of the magnetometer does not need to be known. The total scalar magnetic field is

$$|\mathbf{B}| = \left(|\mathbf{B}_e|^2 + 2\mathbf{B}_e \cdot \Delta \mathbf{B} + |\Delta \mathbf{B}|^2 \right)^{1/2} \tag{4.4}$$

The dipolar field of the Earth is typically 50,000 nT, while the crustal anomalies are much smaller (100–1000 nT). Thus, $|\Delta \mathbf{B}|^2$ is small relative to the other terms, and we can develop an approximate formula for the total scalar field:

$$|\mathbf{B}| \cong |\mathbf{B_e}| \left(1 + \frac{2\Delta \mathbf{B} \cdot \mathbf{B}_e}{|\mathbf{B}_e|^2}\right)^{1/2} \cong |\mathbf{B_e}| \left(1 + \frac{\Delta \mathbf{B} \cdot \mathbf{B}_e}{|\mathbf{B}_e|^2}\right) \tag{4.5}$$

Equation (4.5) can be rearranged to relate the measured scalar anomaly $\mathbf{A}$ to the vector anomaly $\Delta \mathbf{B}$, given an independent measurement of the dipolar field of the Earth $\mathbf{B}_e$:

$$\mathbf{A} = |\mathbf{B}| - |\mathbf{B}_e| = \frac{\Delta \mathbf{B} \cdot \mathbf{B}_e}{|\mathbf{B}_e|} \tag{4.6}$$

4.5 Magnetic Anomalies Due to Seafloor Spreading

To calculate the anomalous scalar field on the sea surface due to thin magnetic stripes on the seafloor, we go back to Poisson's equation relating magnetic field to magnetization. The model is shown in Figure 4.4.

We have an xyz coordinate system with z pointed upward. The $z = 0$ level corresponds to sea level and there is a thin magnetized layer at a depth of z_o.

We define a scalar potential U and a magnetization vector M. The magnetic anomaly $\Delta \mathbf{B}$ is the negative gradient of the potential. The potential satisfies Laplace's equation above the source layer and it satisfies Poisson's equation within the source layer.

$$\Delta \mathbf{B} = -\nabla U \tag{4.7}$$

$$\nabla^2 U = 0 \quad z \neq z_o \tag{4.8}$$

$$\nabla^2 U = \mu_o \nabla \cdot \mathbf{M} \quad z = z_o \tag{4.9}$$

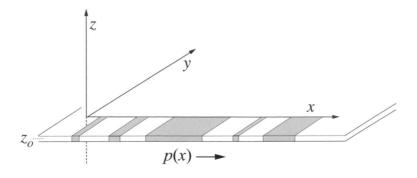

Figure 4.4

$U(x, y, z)$	magnetic potential	T m
μ_o	magnetic permeability	$4\pi \times 10^{-7}$ T A^{-1} m
$\mathbf{M}$	magnetization vector	A m^{-1}

In addition to assuming the layer is infinitesimally thin, we assume that the magnetization direction is constant, but that the magnetization varies in strength and polarity as specified by the reversal function $p(x)$. The approach to the solution is:

1. Solve the differential equation and calculate the magnetic potential U at $z = 0$.
2. Calculate the magnetic anomaly vector $\Delta\mathbf{B}$.
3. Calculate the scalar magnetic field $A = (\Delta\mathbf{B} \cdot \mathbf{B}_e)/|\mathbf{B}_e|$.

Let the magnetization be of the following general form

$$\mathbf{M}(x, y, z) = \left(M_x \hat{i} + M_y \hat{j} + M_z \hat{k} \right) p(x) \delta(z - z_o). \tag{4.10}$$

The differential equation (4.9) becomes

$$\frac{\partial^2 U}{\partial x^2} + \frac{\partial^2 U}{\partial y^2} + \frac{\partial^2 U}{\partial z^2} =$$

$$= \mu_o \left[\frac{\partial}{\partial x} M_x p(x) \delta(z - z_o) + \cancel{\frac{\partial}{\partial y} M_y p(x) \delta(z - z_o)} + \frac{\partial}{\partial z} M_z p(x) \delta(z - z_o) \right].$$

$$\tag{4.11}$$

The y-source term vanishes, because the source does not vary in the y-direction (i.e., the y derivative is zero). Thus the component of magnetization that is parallel to the ridge axis does not produce any external magnetic potential or external magnetic field. Consider a N-S oriented spreading ridge at the magnetic equator. In this case, the TRM of the crust has a component parallel to the dipole field, which happens to be parallel to the ridge axis, so there will be *no external magnetic field anomaly*. See Figure 4.5.

This explains why the global map of magnetic anomaly picks (Cande et al., 1989) has no data in either the equatorial Atlantic or the equatorial Pacific, where ridges are oriented N-S. Now, with the ridge-parallel component of magnetization gone, the differential equation reduces to

$$\frac{\partial^2 U}{\partial x^2} + \frac{\partial^2 U}{\partial z^2} = \mu_o \left[\frac{\partial}{\partial x} M_x p(x) \delta(z - z_o) + \frac{\partial}{\partial z} M_z p(x) \delta(z - z_o) \right]. \tag{4.12}$$

This is a second-order differential equation in two dimensions, so four boundary conditions are needed for a unique solution:

$$\lim_{|x| \to \infty} U(\mathbf{x}) = 0 \quad \text{and} \quad \lim_{|z| \to \infty} U(\mathbf{x}) = 0 \tag{4.13}$$

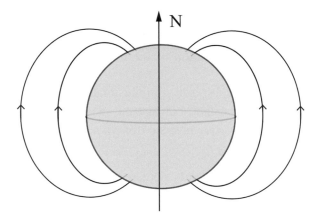

Figure 4.5

Take the two-dimensional Fourier transform of the differential equation where the forward and inverse transforms are defined as

$$F(\mathbf{k}) = \int\limits_{-\infty}^{\infty}\int\limits_{-\infty}^{\infty} f(\mathbf{x})e^{-i2\pi(\mathbf{k}\cdot\mathbf{x})}\mathrm{d}^2\mathbf{x} \quad F(\mathbf{k}) = \Im_2\left[f(\mathbf{x})\right]$$

(4.14)

$$f(\mathbf{x}) = \int\limits_{-\infty}^{\infty}\int\limits_{-\infty}^{\infty} F(\mathbf{k})e^{i2\pi(\mathbf{k}\cdot\mathbf{x})}\mathrm{d}^2\mathbf{k} \quad f(\mathbf{x}) = \Im_2^{-1}\left[F(\mathbf{k})\right]$$

where $\mathbf{x} = (x, z)$ is the position vector, $\mathbf{k} = (k_x, k_z)$ is the wavenumber vector, and $(\mathbf{k} \cdot \mathbf{x}) = k_x x + k_z z$. The derivative property is $\Im_2[\mathrm{d}U/\mathrm{d}x] = i2\pi k_x\,\Im_2[U]$. The Fourier transform of the differential equation is

$$-\left[(2\pi k_x)^2 + (2\pi k_z)^2\right]U(k_x, k_z) = \mu_o p(k_x)e^{-i2\pi k_z z_o}\,(i2\pi\mathbf{k}\cdot\mathbf{M})\,.$$

(4.15)

The Fourier transform in the z-direction was done using the following identity:

$$\int\limits_{-\infty}^{\infty} \delta\,(z - z_o)e^{-i2\pi k_z z}\,\mathrm{d}z \equiv e^{-i2\pi k_z z_o}$$

(4.16)

Now we can solve for $U(\mathbf{k})$:

$$U(\mathbf{k}) = \frac{-i\mu_o}{2\pi}p(k_x)(\mathbf{k}\cdot\mathbf{M})\frac{e^{-i2\pi k_z z_o}}{\left(k_x^2 + k_z^2\right)}$$

(4.17)

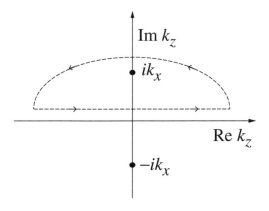

Figure 4.6

Next, take the inverse Fourier transform with respect to k_z, using the Cauchy residue theorem:

$$U(k_x, z) = \frac{\mu_o}{2\pi i} p(k_x) \int_{-\infty}^{\infty} \frac{(\mathbf{k} \cdot \mathbf{M}) e^{i2\pi k_z(z-z_o)}}{\left(k_x^2 + k_z^2\right)} dk_z \qquad (4.18)$$

The poles of the integrand are found by factoring the denominator:

$$k_x^2 + k_z^2 = (k_z + ik_x)(k_z - ik_x) \qquad (4.19)$$

We see that $U(\mathbf{k})$ is an analytic function with poles at $\pm ik_x$. The integral of this function about any closed path in the complex k_z plane is zero, unless the contour includes a pole, in which case the integral is $i2\pi$ times the residue at the pole:

$$\oint \frac{f(z)}{z - z_o} dz = i2\pi f(z_o) \qquad (4.20)$$

One possible path integral is shown in Figure 4.6.

There are two ways to close the path at infinity. The selection of the proper path—and thus the residue—depends on the boundary condition, equation (4.13). First, consider the case where $k_x > 0$. If we close the path of integration in the upper imaginary plane, then the pole will be ik_x. The residue will have an exponential term that vanishes as z goes to plus infinity. This is what we need, since the observation plane is above the source.

$$\oint () \, dk_z = \frac{e^{-2\pi k_x(z-z_o)}}{2ik_x} (k_x M_x + ik_x M_z) \qquad (4.21)$$

Next, consider the case where $k_x < 0$. To satisfy the boundary condition as z goes to plus infinity, the $-ik_x$ pole should be used, and the integration path will be clockwise instead of counterclockwise, as in equation (4.20).

$$\oint () \, dk_z = \frac{e^{+2\pi k_x (z-z_0)}}{2i k_x} (k_x M_x - i k_x M_z) \tag{4.22}$$

One can combine the two cases by using the absolute value of k_x:

$$U(k,z) = \frac{\mu_o}{2} p(k) e^{-2\pi |k| (z-z_0)} \left(M_z - i \frac{k}{|k|} M_x \right) \tag{4.23}$$

where we have dropped the subscript on the x-wavenumber.

This is the general case of an infinitely long ridge. To further simplify the problem, let's assume that this spreading ridge is located at the magnetic pole of the Earth, so the dipolar field lines will be parallel to the z-axis and there will be no x-component of magnetization. The result is

$$U(k,z) = \frac{\mu_o M_z}{2} p(k) e^{-2\pi |k| (z-z_o)} \tag{4.24}$$

Next calculate the magnetic anomaly $\Delta B = -\nabla U$:

$$\Delta \mathbf{B} = (-i 2\pi k, 2\pi |k|) \, U(k,z) \tag{4.25}$$

The scalar magnetic field is given by equation (4.6). Since only the z-component of the Earth's field is non-zero, the anomaly simplifies to

$$\underbrace{A(k,z)}_{\substack{\text{observed} \\ \text{anomaly}}} = \frac{\mu_o M_z}{2} \underbrace{p(k)}_{\substack{\text{reversal} \\ \text{pattern}}} \times \underbrace{2\pi |k| \, e^{-2\pi |k| (z-z_o)}}_{\substack{\text{Earth} \\ \text{filter}}}. \tag{4.26}$$

The reversal pattern is a sequence of positive and negative polarities. To generate the model anomaly, one would take the Fourier transform of the reversal pattern, multiply by the Earth filter, and take the inverse transform of the result. An examination of the Earth filter in Figure 4.7 illustrates why a square-wave reversal pattern becomes distorted.

This Earth filter attenuates both long and short wavelengths, so it acts like a band-pass filter. In the space domain it modifies the shape of the square-wave reversal pattern, as shown in Figure 4.8.

When the seafloor spreading ridge is not at the magnetic pole, both the magnetization and the Earth's magnetic field will have an x-component. This introduces a

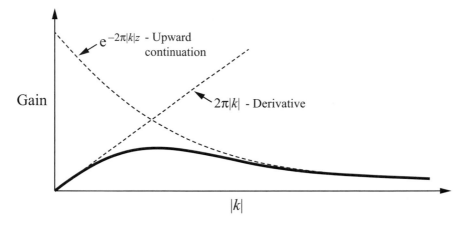

Figure 4.7 The Earth filter is the product of a derivative filter and an upward continuation filter.

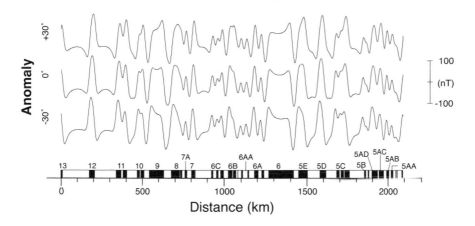

Figure 4.8 Synthetic magnetic anomalies generated from the reversal pattern (lower) using a 0.5 km thick magnetized layer at a depth of 4.25 km and a spreading rate of 100 mm/yr. The three curves have skewness of $+30, 0$, and -30 degrees. From Horner-Johnson and Gordon (2003).

phase shift, or skewness Θ, in the output magnetic anomaly. At the ocean surface, the skewed magnetic anomaly is

$$A(k) = \frac{\mu_o M_z}{2} p(k) e^{i\Theta \frac{k}{|k|}} 2\pi |k| e^{+2\pi |k| z_o}. \tag{4.27}$$

The skewness depends on both the geomagnetic latitude and the orientation of the spreading ridge when the crust was magnetized. Moreover, this parameter will vary over time. If one knows the skewness, then the model profile can be skewed to match the observed profile. Alternately, the observed magnetic anomaly can be de-skewed. This is called *reduction to the pole*, because it synthesizes the anomaly that would have formed on the magnetic pole.

$$A_{pole}(k) = A_{observed}(k)e^{-i\Theta \frac{k}{|k|}} \tag{4.28}$$

4.6 Discussion

The ability to observe magnetic reversals from a magnetometer towed behind a ship relies on some remarkable coincidences related to reversal rate, spreading rate, ocean depth, and Earth temperatures (mantle, seafloor, and Curie). In the case of marine magnetic anomalies, four scales must match.

First, the temperature of the mantle (1200 °C), the seafloor (0 °C), and the Curie temperature of basalt ($\sim$500 °C) must be just right for recording the direction of the Earth's magnetic field at the seafloor spreading ridge axis. Most of the thermoremnant magnetism (TRM) is recorded in the upper 1000 m of the oceanic crust. If the thickness of the TRM layer was too great, then as the plate cooled while it moved off the spreading ridge axis, the positive and negative reversals would be juxtaposed in dipping vertical layers (Figure 4.3). This superposition would smear the pattern observed by a ship. If the seafloor temperature was above the Curie temperature, as it is on Venus, then no recording would be possible.

The second scale is related to ocean floor depth and thus the Earth filter. The external magnetic field is the derivative of the magnetization, which, as shown above, acts as a high-pass filter applied to the reversal pattern recorded in the crust. The magnetic field measured at the ocean surface will be naturally smooth (upward continuation), due to the distance from the seafloor to the sea surface; this is a low-pass filter. This smoothing depends exponentially on ocean depth, so for a wavelength of 2π times the mean ocean depth, the field amplitude will be attenuated by $1/e$, or 0.37, with respect to the value measured at the seafloor. The combined result of the derivative and the upward continuation is a band-pass filter with a peak response at a wavelength of 2π times the mean ocean depth, or about 25 km. Wavelengths that are shorter ($<$10 km) or much longer ($>$500 km) than this value will be undetectable at the ocean surface.

The third and fourth scales that must match are the reversal rate and the seafloor-spreading rate. Half-spreading rates on the Earth vary from 10 to 80 km per million years. Thus, for the magnetic anomalies to be most visible on the ocean surface, the reversal rate should be between 2.5 and 0.3 million years. It is astonishing that this is the typical reversal rate observed in sequences of lava flows on land! While most ocean basins display clear reversal patterns, there was a period between 85 and 120 million years ago when the magnetic field polarity of the Earth remained positive, so the ocean surface anomaly is too far from the reversal boundaries to provide timing information. This area of seafloor is called the Cretaceous quiet zone; it is a problem area for accurate plate reconstructions.

The lucky convergence of length and timescales makes it very unlikely that magnetic anomalies due to crustal spreading will ever be observed on another planet.

4.7 Exercises

Exercise 4.1 Explain why magnetic lineations cannot be observed from a spacecraft orbiting the Earth at an altitude of 400 km.

Exercise 4.2 Explain why scalar magnetic anomalies are not observed at a N S oriented spreading ridge located at the magnetic equator.

Exercise 4.3 Write a MATLAB program to generate marine magnetic anomaly versus distance from a spreading ridge axis. Use equation (4.27) relating the Fourier transform of the magnetic anomaly to the Fourier transform of the magnetic timescale. You will need a magnetic timescale and the start of a MATLAB program (topex .ucsd.edu/pub/class/geodynamics/hw3). Assume symmetric spreading about the ridge axis, constant spreading rate, and constant ocean depth.

Use the program and magnetic anomaly profiles across the Pacific-Antarctic Rise (NBP9707.xydm) and the Mid-Atlantic Ridge (a9321.xydm) to estimate the half-spreading rate at each of these ridges. You may need to vary the mean ocean depth and skewness to obtain good fits.

Describe some of the problems that you had fitting the data. Provide some estimates on the range of total spreading rate for each ridge.

Exercise 4.4 Explain how the global gridded data set EMAG2 was constructed (www.ngdc.noaa.gov/geomag/emag2.html). Download the grid as a geotiff file and extract a subgrid approximately 2000 km by 2000 km. Use the upward continuation formula $A(\mathbf{k}, z) = A(\mathbf{k}, 0)e^{-2\pi|\mathbf{k}|z}$ to calculate the magnetic field at an altitude of 450 km. Explain why satellite measurements of the magnetic field cannot be used to map ocean anomalies related to seafloor spreading.

5

Cooling of the Oceanic Lithosphere

5.1 Introduction

This chapter uses the Fourier transform method to solve for the temperature in the cooling oceanic lithosphere. For researchers in the areas of marine geology, marine geophysics, and geodynamics, this is the most important concept you can learn from this book. As noted in the original paper on the topic by Turcotte and Oxburgh (1967), convection of the mantle is primarily controlled by thin thermal boundary layers. The surface thermal boundary layer, or *oceanic lithosphere*, is the most important component of the convecting system, because it represents the greatest temperature gradient in the Earth. It also has a greater surface area than the second-most important thermal boundary layer, which is at the core–mantle boundary. As the lithosphere cools it becomes denser, the seafloor depth increases, and ultimately the lithosphere founders (*subduction*). This subduction process both drives the convective flow and efficiently quenches the mantle.

This chapter covers the same material as *Geodynamics* (Turcotte and Schubert, 2014, Sections 4.15 to 4.17). The main difference is the method of solution. Turcotte and Schubert solve the half-space cooling problem by using a similarity variable to reduce the time-dependent heat conduction equation from a partial differential equation to an ordinary differential equation that can be solved by integration. These notes provide an alternate solution to the problem by using the tools of Fourier analysis. Basically, any type of heat conduction problem can be solved with the Fourier approach (Carslaw and Jaeger, 1959). This Fourier approach is more than just a new way to solve an old problem. Many 3-D heat conduction problems with complicated sources and boundary conditions do not have complete analytic solutions, but do have solutions in the Fourier transform domain. In these cases, the FFT algorithms, coupled with modern computers, can be used to compute accurate results in seconds. Resorting to finite difference or other

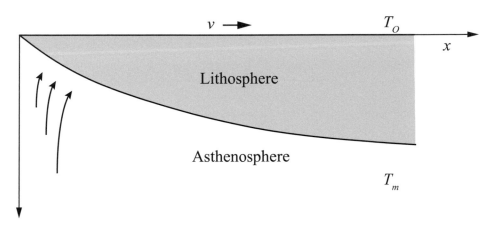

Figure 5.1

numerical schemes is error-prone and the results are more difficult to interpret, since the analytic foundation is gone. Thus, the Fourier approach is worth learning.

The basic model is shown in Figure 5.1, which represents one half of a seafloor spreading system.

The model assumptions and consequences are:

- lithospheric plates are rigid and move away from the spreading ridge axis at a uniform rate of v;
- hot, low-viscosity asthenosphere fills the void (passive);
- internal heat generation is much smaller than the other terms in the heat equation, so it is neglected; and
- there is a singular point at $x = z = 0$. (This heat is released by hydrothermal circulation.)

This is a two-dimensional problem with no heat sources, so the heat equation has only diffusive and advective terms

$$\frac{\partial^2 T}{\partial x^2} + \frac{\partial^2 T}{\partial z^2} = \frac{v}{\kappa}\frac{\partial T}{\partial x} \tag{5.1}$$

where T is temperature and κ is the thermal diffusivity. The first term represents the lateral diffusion of heat, the second term represents the vertical diffusion of heat, and the third term (on the right side) is the advection of heat by the motion of the plate. Away from the ridge axis, the lateral heat diffusion is much smaller than the vertical heat diffusion. Dropping this term simplifies the differential equation, although a solution can also be developed where the term is retained. Next we move from a Eulerian coordinate system to a Lagrangian system moving with the lithosphere.

$$v = \frac{\partial x}{\partial t} \rightarrow \frac{\partial T}{\partial x}\frac{\partial x}{\partial t} = \frac{\partial T}{\partial t} \qquad (5.2)$$

This reduces the problem to the half-space cooling problem.

$$\frac{\partial^2 T}{\partial z^2} = \frac{1}{\kappa}\frac{\partial T}{\partial t} \qquad (5.3)$$

The boundary and initial conditions are

$$T(0,t) = T_o$$

$$T(\infty,t) = T_m \qquad (5.4)$$

$$T(z,0) = T_m.$$

The infinite half-space has constant thermal diffusivity and an initially constant temperature T_m. At times greater than zero, the surface temperature is T_o. The temperature will evolve with time. Note that for this problem, time also corresponds to the age of the cooling oceanic lithosphere. Define a dimensionless temperature as

$$\theta = \frac{T - T_o}{T_m - T_o}. \qquad (5.5)$$

Now the differential equation and boundary conditions become

$$\frac{\partial^2 \theta}{\partial z^2} = \frac{1}{\kappa}\frac{\partial \theta}{\partial t}$$

$$\theta(0,t) = 0 \qquad (5.6)$$

$$\theta(\infty,t) = 1$$

$$\theta(z,0) = 1.$$

Turcotte and Schubert (2014, page 184) introduce the following dimensionless quantity and use it to reduce equation (5.6) to an ordinary differential equation with two boundary conditions.

$$\eta = \frac{z}{2\sqrt{\kappa t}} \qquad (5.7)$$

They then integrate the differential equation twice and match the boundary conditions. Suppose one did not know this trick or the problem was more complicated.

An approach called *method of images* is straightforward. The model is expanded to a full-space with an initial step-function temperature distribution, so the zero-temperature boundary condition is always matched. The problem becomes

$$\frac{\partial^2 \theta}{\partial z^2} = \frac{1}{\kappa} \frac{\partial \theta}{\partial t}$$

$$\theta(\infty, t) = 1 \tag{5.8}$$

$$\theta(z, 0) = 2H(z) - 1$$

where the definition of the step function is

$$H(z) \equiv \int_{-\infty}^{z} \delta(\xi) \, d\xi. \tag{5.9}$$

Now take the Fourier transform of equation (5.8) with respect to z. The differential equation becomes

$$-\kappa (2\pi k)^2 \Theta(k, t) = \frac{\partial \Theta}{\partial t}. \tag{5.10}$$

The general solution is

$$\Theta(k, t) = C_o e^{-\kappa(2\pi k)^2 t}. \tag{5.11}$$

Now take the Fourier transform of the initial condition.

$$\Im[\Theta(k, 0)] = \Im[2H(z)] - \Im[1] \tag{5.12}$$

We know that

$$\Im[1] = \delta(k). \tag{5.13}$$

Also, using the derivative property we know that

$$\Im\left[\frac{\partial H}{\partial z}\right] = i2\pi k \, \Im[H(z)]. \tag{5.14}$$

Since the derivative of the step function is a delta function, the Fourier transform of the initial condition is

$$\Theta(k, 0) = \frac{1}{i\pi k} - \delta(k). \tag{5.15}$$

The solution that satisfies the initial condition is

$$\Theta(k,t) = \left[\frac{1}{i\pi k} - \delta(k)\right]e^{-\kappa(2\pi k)^2 t}. \tag{5.16}$$

Now we take the inverse Fourier transform.

$$\theta(z,t) = \int_{-\infty}^{\infty} \frac{e^{-\kappa(2\pi k)^2 t}}{i\pi k} e^{i2\pi kz}\,dk - \int_{-\infty}^{\infty} \delta(k)\,e^{-\kappa(2\pi k)^2 t}\,e^{i2\pi kz}\,dk \tag{5.17}$$

The second integral on the right side of equation (5.17) is equal to 1, since the delta function extracts the integrand at $k = 0$. The first integral on the right side of equation (5.17) is performed in two steps. First take the derivative with respect to z to note that

$$\frac{\partial\theta(z,t)}{\partial z} = 2\int_{-\infty}^{\infty} e^{-\kappa(2\pi k)^2 t} e^{i2\pi kz}\,dk. \tag{5.18}$$

This is the Fourier transform of a Gaussian function. The following substitution puts the integral in the form that appears in Bracewell (1978).

$$k' = k\sqrt{4\pi\kappa t} \quad\text{and}\quad z' = \frac{z}{\sqrt{4\pi\kappa t}} \tag{5.19}$$

The result is

$$\frac{\partial\theta(z,t)}{\partial z} = \frac{2}{\sqrt{4\pi\kappa t}}e^{\frac{-z^2}{4\kappa t}}. \tag{5.20}$$

Next integrate equation (5.20) over z. The introduction of the similarity variable based on equation (5.20) helps to identify the integral as the definition of the error function.

$$\eta = \frac{z}{2\sqrt{\kappa t}} \quad\text{so}\quad dz = 2\sqrt{\kappa t}\,d\eta \tag{5.21}$$

The integral becomes

$$\theta(z,t) = \frac{2}{\sqrt{\pi}}\int_{\infty}^{\eta} e^{-\eta^2}\,d\eta - 1. \tag{5.22}$$

The right side of equation (5.22) is just the definition of the error function erf(η). The final solution is

$$T(z,t) = (T_m - T_o)\,\text{erf}\left(\frac{z}{2\sqrt{\kappa t}}\right) + T_o. \tag{5.23}$$

5.2 Temperature versus Depth and Age

The thermal parameters and temperatures appropriate to the Earth are given in Table 5.1.

If we define the base of the thermal boundary layer as some large fraction of the deep mantle temperature, as in the table, one can calculate the thickness of the thermal boundary layer versus the age of the lithosphere.

$$\frac{T_l - T_o}{T_m - T_o} = 0.84 = \text{erf}\left(\frac{z}{2\sqrt{\kappa t}}\right) \tag{5.24}$$

or

$$z \cong 2\sqrt{\kappa t} \quad \text{or} \quad z(\text{km}) \cong 10\sqrt{\text{age}(\text{Ma})}. \tag{5.25}$$

The isotherms for this model are displayed in Figure 5.2.

Table 5.1.

Parameter	Definition	Value
T_o	surface temperature	$0\,^\circ\text{C}$
T_l	temperature at base of thermal boundary layer	$1100\,^\circ\text{C}$
T_m	mantle temperature	$1300\,^\circ\text{C}$
κ	thermal diffusivity	$8 \times 10^{-7}\ \text{m}^2\,\text{s}^{-1}$
k	thermal conductivity	$3.3\ \text{W m}^{-1}\,^\circ\text{C}^{-1}$

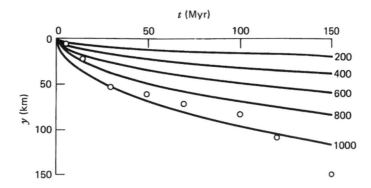

Figure 5.2 The solid lines are isotherms in the oceanic lithosphere from equation (5.23). The data points are the thickness of the oceanic lithosphere in the Pacific determined from studies of Rayleigh wave dispersion data. (From Leeds and Kausel (1974).)

5.3 Heat Flow versus Age

The heat flow is the thermal conductivity times the temperature gradient.

$$q(z) = k \frac{\partial T}{\partial z} \tag{5.26}$$

To calculate the heat flow we take the derivative of the error function with respect to z.

$$\frac{\partial \, \text{erf}(\eta)}{\partial z} = \frac{\partial \, \text{erf}(\eta)}{\partial \eta} \frac{\partial \eta}{\partial z} = \frac{1}{\sqrt{\pi \kappa t}} e^{-\eta^2} \tag{5.27}$$

$$q(z,t) = \frac{k(T_m - T_o)}{\sqrt{\pi \kappa t}} e^{\frac{-z^2}{4\kappa t}} \tag{5.28}$$

In the limit as depth z goes to infinity, the heat flow is zero. So for this model, there is no heat transport into the base of the lithosphere. Later we'll compute seafloor depth versus age for this model and show that there are large deviations at old age (i.e., >70 Ma). One way to flatten the depth-versus-age curve is to supply heat to the base of the lithosphere. There are a variety of ways to accomplish this.

- Increasing basal heat flux with age corresponds to the plate cooling model of Parsons and Sclater (1977). The physical mechanism for this basal heat input is small-scale convective rolls beneath the old lithosphere.
- A constant basal heat flux with age corresponds to the CHABLIS cooling model of Doin and Fleitout (1996).
- Some papers (e.g., Crough (1983)) propose that mantle plumes reheat the old lithosphere and eventually all old lithosphere encounters one or more plumes, so reheating is pervasive.

The surface heat flow is just equation (5.28) evaluated at the surface of the earth.

$$q(t) = \frac{k(T_m - T_o)}{\sqrt{\pi \kappa t}} \tag{5.29}$$

The match to the observed heat flow is shown in Figure 5.3. For ages less than about 40 Ma, the surface heat flux is less than predicted by the model. This heat flow deficit occurs because cold seawater circulates deep into the crust and advects the heat. So the temperature gradient will be less than predicted by a purely conductive model. At older ages, the heat flow is higher than expected. This could either be due to a non-zero basal heat flux or an incorrect estimate of thermal conductivity of the crust.

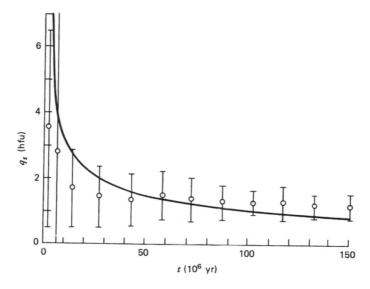

Figure 5.3 Mean values and standard deviations of ocean floor heat flow measurements as functions of age compared with equation (5.29). Data from Sclater et al. (1980).

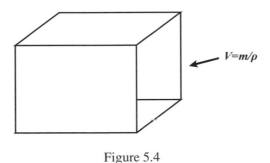

Figure 5.4

5.4 Thermal Subsidence

As the oceanic lithosphere cools by conductive heat loss, it contracts. This thermal contraction causes the average density of the lithosphere to increase. The seafloor depth increases with age and eventually the lithosphere becomes so dense it founders at a subduction zone. To develop a linear relationship between density and temperature, consider a cube of volume V, mass m, and density ρ, at temperature T_o, under a confining pressure P_o. (See Figure 5.4.)

Changes in both temperature and pressure will produce changes in the volume of the cube.

$$dV = \left(\frac{\partial V}{\partial T}\right)_{P_o} dT + \left(\frac{\partial V}{\partial P}\right)_{T_0} dP \tag{5.30}$$

The two terms in equation (5.30) are related to the volumetric coefficient of thermal expansion,

$$\alpha = \frac{1}{V}\left(\frac{\partial V}{\partial T}\right)_{P_o}$$

(5.31)

and the isothermal compressibility is

$$\beta = -\frac{1}{V}\left(\frac{\partial V}{\partial P}\right)_{T_o}.$$

(5.32)

Since $\rho = mV^{-1}$ it is easy to show that

$$\frac{\partial \rho}{\rho} = -\frac{\partial V}{V},$$

(5.33)

so the coefficient of thermal expansion becomes

$$\alpha = -\frac{1}{\rho}\left(\frac{\partial \rho}{\partial T}\right)_{P_o}.$$

(5.34)

In this model, the lithosphere slides laterally across the surface of the earth, so there are no significant pressure variations. Thus, we need only the first term in equation (5.30). If ρ_m is the density of the lithosphere at a temperature of T_m, then a reduction in temperature will cause an increase in density.

$$\rho(T) = \rho_m\left[1 - \alpha(T - T_m)\right]$$

(5.35)

The diagram in Figure 5.5 illustrates the thermal subsidence of the oceanic lithosphere as it spreads from the ridge axis at a velocity of v.

There are three layers in the model. The ocean has a density of ρ_w and a depth of d_0 at the ridge axis. This depth increases with age/distance from the ridge axis. We

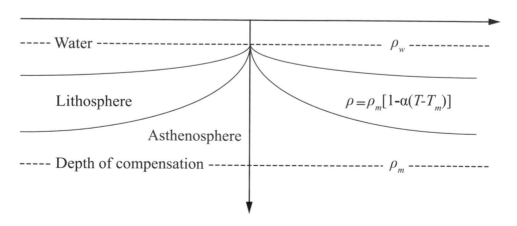

Figure 5.5

will use the principles of thermal contraction and isostasy to determine the increase in seafloor depth with increasing age $d(t)$. The density of the lithosphere depends on temperature, according to equation (5.35). The asthenosphere behaves as a fluid on geological timescales, so the lithosphere floats on the mantle.

The major assumptions are:

- The pressure at the depth of compensation is a constant value and depends only on the weight of the rock and water directly above (i.e., isostatic equilibrium).
- The crust has uniform thickness, so it has no effect on the overall isostatic balance.
- The thermal diffusivity κ is isotropic and independent of P and T.
- The thermal expansion coefficient α is isotropic and independent of P and T.
- Heat is transferred by conduction, so hydrothermal circulation is not important. This is a poor assumption at the ridge axis.
- Heat conducts only vertically. This is also a poor assumption at the ridge axis.
- There are no heat sources in the crust or lithosphere.
- No heat flows into the base of the lithosphere See Doin and Fleitout (1996) for a discussion of alternate models with basal heat input.

An additional assumption is that the lithosphere is free to contract in all three dimensions. Since the lithosphere is thin in relation to its horizontal dimension, free contraction in the vertical dimension is a good assumption. Contraction of the plate in the direction perpendicular to the ridge axis is probably valid as well. However, contraction in the ridge-parallel direction will produce significant shear strain, which will result in thermoelastic stress. We will neglect this for now but this is an interesting area of research.

As the lithosphere cools and contracts, its vertically integrated density increases, which will increase the pressure at its base. To maintain isostatic balance (i.e., constant pressure at constant depth z_l), ocean depth must increase to replace high density rock with lower density water. The increase in depth is determined by the following isostatic balance between a ridge-axis column and an off-axis column. See Figure 5.6.

The mathematical statement of isostatic balance is

$$g \int_{o}^{z_l} \rho_m \, dz = g \int_{o}^{d} \rho_w \, dz + g \int_{d}^{z_l} \rho_m \left[1 - \alpha \left(T - T_m \right) \right] dz \qquad (5.36)$$

where g is the acceleration of gravity.

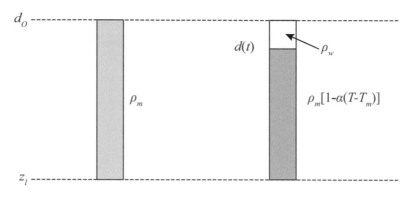

Figure 5.6

After subtracting the standard ridge-axis column from both sides and dividing through by g we get

$$0 = \int_o^d (\rho_w - \rho_m)\, dz - \int_d^{z_l} \rho_m\, \alpha\left(T - T_m\right) dz. \tag{5.37}$$

Now we'll use the solution to the half-space cooling problem (equation (5.23)) to define $T(t, z)$. Note this solution has temperature perturbations at infinite depth, so we must extend the depth integration from the seafloor to infinity.

$$d\left(\rho_m - \rho_w\right) = \rho_m\, \alpha\left(T_m - T_o\right) \int_d^{\infty} 1 - \operatorname{erf}\left(\frac{z - d}{2\sqrt{\kappa t}}\right) dz \tag{5.38}$$

By setting $z' = z - d$ and solving for $d(t)$, we find

$$d(t) = \frac{\rho_m\, \alpha\left(T_m - T_o\right)}{\left(\rho_m - \rho_w\right)} \int_o^{\infty} \operatorname{erfc}\left(\frac{z}{2\sqrt{\kappa t}}\right) dz. \tag{5.39}$$

To integrate this function, let $\eta = z\big/\left(2\sqrt{\kappa t}\right)$, so $dz = 2\sqrt{\kappa t}\, d\eta$

$$d(t) = \frac{2\rho_m\, \alpha\left(T_m - T_o\right)}{\left(\rho_m - \rho_w\right)} \sqrt{\kappa t} \int_o^{\infty} \operatorname{erfc}\left(\eta\right) d\eta. \tag{5.40}$$

After performing the definite integral of

$$\int_o^{\infty} \operatorname{erfc}\left(\eta\right) d\eta = \frac{1}{\sqrt{\pi}}$$

Table 5.2.

Parameter	Definition	Value
T_o	surface temperature	$0\,°C$
T_m	mantle temperature	$1365\,°C$
κ	thermal diffusivity	$8 \times 10^{-7}\,m^2\,s^{-1}$
k	thermal conductivity	$3.3\,W\,m^{-1}\,°C^{-1}$
α	thermal expansion coefficient	$3.1 \times 10^{-5}\,°C^{-1}$
ρ_w	seawater density	$1025\,kg\,m^{-3}$
ρ_m	mantle density	$3300\,kg\,m^{-3}$
d_o	ridge axis depth	$2500\,m$
L	asymptotic plate thickness	$125\,km$

and adding the ridge axis depth d_o, we find that depth depends on material constants times the square root of seafloor age.

$$d_{\text{tot}}(t) = d_o + \frac{2\rho_m\,\alpha\,(T_m - T_o)}{(\rho_m - \rho_w)}\left(\frac{\kappa t}{\pi}\right)^{1/2} \tag{5.41}$$

Now let's plug in some numbers to get an estimate of how seafloor depth varies with age (Table 5.2).

A good approximation for the depth-age relation is

$$d(\text{m}) = 2500 + 350\sqrt{\text{age(Ma)}}. \tag{5.42}$$

To test this model of the cooling oceanic lithosphere, we need seafloor depth, seafloor age, and sediment thickness (Renkin and Sclater, 1988).

5.5 The Plate Cooling Model

The half-space cooling model developed above provides a remarkably accurate description of the variations in heat flow and depth versus the age of the seafloor, for ages less than about 70 Ma. However, for older ages there is a pronounced flattening of the seafloor depth that is better fit by the plate cooling model (Parsons and Sclater, 1977). In this section we develop formulas for the temperature, heat flow, and depth for the plate model and show comparisons with heat flow and depth data. Later we will use the same model to investigate the thickness and strength of the cooling oceanic lithosphere as well as the major driving forces for plate tectonics.

We begin with the 1-D heat diffusion equation

$$\frac{\partial^2 T}{\partial z^2} = \frac{1}{\kappa}\frac{\partial T}{\partial t}.$$ (5.43)

The only difference between the plate model and the half-space cooling model is that the plate model has a lithosphere of finite thickness L. The initial condition and boundary conditions are

$$T(0,t) = T_o$$
$$T(L,t) = T_m$$ (5.44)
$$T(z,0) = T_m.$$

As in the case of the half-space cooling derivation, we non-dimensionalize the temperature.

$$\theta' = \frac{T - T_o}{T_m - T_o}$$ (5.45)

In addition, we recognize that as $t \to \infty$, the temperature increases linearly with depth so we can define the long-term non-dimensional temperature as

$$\theta' = \theta(z,t) + \frac{z}{L}$$ (5.46)

where $\theta(z,t)$ is the transient part of the solution that goes to zero at large time. The differential equation and boundary conditions become

$$\frac{\partial^2 \theta}{\partial z^2} = \frac{1}{\kappa}\frac{\partial \theta}{\partial t}$$
$$\theta(0,t) = 0$$
$$\theta(L,t) = 0$$ (5.47)
$$\theta(z,0) = 1 - \frac{z}{L}; z > 0.$$

We will use separation of variables to decompose the solution as the product of two functions.

$$\theta(z,t) = g(z)\,f(t)$$ (5.48)

Note that a Fourier sine series in depth z automatically satisfies the zero boundary temperature at the top and bottom of the plate. We guess the form of the solution as

$$\theta(z,t) = \sum_{n-1}^{\infty} a_n \sin\left(\frac{n\pi z}{L}\right) f_n(t)$$ (5.49)

where $f(0) = 1$. The Fourier coefficients are given by

$$a_n = \frac{2}{L} \int_0^L \theta(z,0) \sin\left(\frac{n\pi z}{L}\right) dz. \tag{5.50}$$

The differential equations should be satisfied for each value of n as

$$\frac{\partial^2 \theta_n}{\partial z^2} = \frac{1}{\kappa} \frac{\partial \theta_n}{\partial t}. \tag{5.51}$$

After a little algebra this becomes

$$-a_n \kappa \left(\frac{n\pi}{L}\right)^2 \sin\left(\frac{n\pi z}{L}\right) f_n(t) = a_n \sin\left(\frac{n\pi z}{L}\right) \frac{\partial f_n(t)}{\partial t} \tag{5.52}$$

or

$$-\kappa \left(\frac{n\pi}{L}\right)^2 f_n(t) = \frac{\partial f_n(t)}{\partial t}. \tag{5.53}$$

The time-dependent solution that satisfies the boundary conditioni $f(0) = 1$ is

$$f_n(t) = e^{-\kappa \left(\frac{n\pi}{L}\right)^2 t} \tag{5.54}$$

and the time-dependent solution is

$$\theta_n(z,t) = a_n \sin\left(\frac{n\pi z}{L}\right) e^{-\kappa \left(\frac{n\pi}{L}\right)^2 t}. \tag{5.55}$$

Finally, we need to determine the Fourier sine coefficients by inserting the initial temperature distribution $1 - \frac{z}{L}$ into equation (5.50) and performing the integration. This is left as an exercise to show that $a_n = \frac{2}{n\pi}$.

The final result is

$$\theta(z,t) = \frac{2}{\pi} \sum_{n=1}^{\infty} \frac{1}{n} \sin\left(\frac{n\pi z}{L}\right) e^{-\kappa \left(\frac{n\pi}{L}\right)^2 t} \tag{5.56}$$

so the total solution is

$$\theta_n'(z,t) = \frac{z}{L} + \frac{2}{\pi} \sum_{n=1}^{\infty} \frac{1}{n} \sin\left(\frac{n\pi z}{L}\right) e^{-\kappa \left(\frac{n\pi}{L}\right)^2 t}. \tag{5.57}$$

The final temperature is given by

$$T(z,t) = T_o + (T_m - T_o) \left[\frac{z}{L} + \frac{2}{\pi} \sum_{n=1}^{\infty} \frac{1}{n} \sin\left(\frac{n\pi z}{L}\right) e^{-\kappa \left(\frac{n\pi}{L}\right)^2 t} \right]. \tag{5.58}$$

From this we can calculate the heat flow

$$q = -k\frac{\partial T}{\partial z} = \frac{k\,(T_m - T_o)}{L}\left[1 + 2\sum_{n=1}^{\infty}\cos\left(\frac{n\pi z}{L}\right)e^{-\kappa\left(\frac{n\pi}{L}\right)^2 t}\right]. \tag{5.59}$$

The heat flow at the surface and the bottom of the plate is

$$q(0,t) = \frac{k\,(T_m - T_o)}{L}\left[1 + 2\sum_{n=1}^{\infty}e^{-\kappa\left(\frac{n\pi}{L}\right)^2 t}\right]$$

$$q(L,t) = \frac{k\,(T_m - T_o)}{L}\left[1 + 2\sum_{n=1}^{\infty}(-1)^n e^{-\kappa\left(\frac{n\pi}{L}\right)^2 t}\right]. \tag{5.60}$$

In Section 5.4 on thermal subsidence we used the principle of isostatic compensation to calculate the increase in seafloor depth as a function of cooling time or age. The general formula is given in 5.37 and it is repeated here

$$d\,(\rho_m - \rho_w) = \int_0^L \rho_m\alpha\,(T_m - T)\mathrm{d}z \tag{5.61}$$

Using equation 5.58 this becomes

$$d\,(t)\,(\rho_m - \rho_w) = \alpha\rho_m\,(T_m - T_o)\int_0^L\left[1 - \frac{z}{L} - \frac{2}{\pi}\sum_{n=1}^{\infty}\frac{1}{n}\sin\left(\frac{n\pi z}{L}\right)e^{-\kappa\left(\frac{n\pi}{L}\right)^2 t}\right] \tag{5.62}$$

It is left as an exercise to show the final result is

$$d(t) = d_o + \frac{\alpha\rho_m\,(T_m - T_o)\,L}{2\,(\rho_m - \rho_w)}\left[1 - \frac{8}{\pi^2}\sum_{n=1}^{\infty}\frac{1}{(2n-1)^2}e^{-\kappa\left(\frac{(2n-1)\pi}{L}\right)^2 t}\right] \tag{5.63}$$

where we have added the constant depth at the ridge axis d_o.

The plate cooling model provides a good fit to the heat and depth versus age data as shown in Figure 5.7. The upper plot shows contours of temperature versus depth and age. The solid curves are for the plate model with a thickness of 125 km while the dashed curves are for the half space model. The lower constant temperature boundary condition of the plate model is maintained by an increase in heat flow

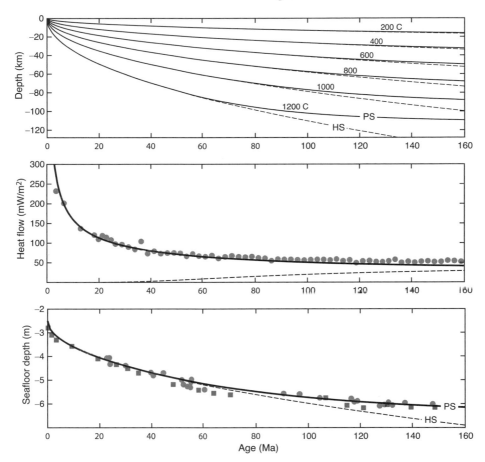

Figure 5.7 (upper) Temperature contours as a function of depth and age for the plate (PS – solid curves) (Parsons and Sclater, 1977) and half-space (HS – dashed curves) cooling models. Flattening of the 1200 °C isotherm begins at an age of 70 Ma. (middle) Heat flow for plate cooling model (solid curve) and heat flow data (circles) from Hasterok (2013). (lower) Depth versus age for the plate and half space cooling models. Data are from Parsons and Sclater (1977) (boxes) and Garcia et al. (2019) (circles).

into the base of the lithosphere as shown in the dashed line of the heat flow plot (middle). These heat flow data from Hasterok (2013) were carefully selected to avoid areas of thin sediment cover where hydrothermal circulation extracts heat from the crust and lowers the conductive heat flow. The depth versus age data were also carefully processed to avoid areas of thickened crust and account for the thickness of the sediments. Both effects will make the depth appear shallower which has been interpreted by some authors (Stein and Stein, 1992) as evidence for a thinner lithosphere (95 km instead of 125 km). When we discuss flexure and the yield strength of the oceanic lithosphere, we will see that the thicker lithosphere is required to support the trench and outer rise topography.

5.6 Buoyancy of the Cooling Lithosphere

Cooling and contraction of the lithosphere causes seafloor depth to increase with age. The average density of cooled mantle lithosphere is greater than the density of the underlying mantle, so it would founder and subduct if it could be decoupled from more buoyant surrounding lithosphere (e.g., continents). Once subduction begins, the negative buoyancy of the subducted slab pulls the surface plate into the mantle driving plate tectonics. The amount of slab pull is related to the average density of the lithosphere relative to the mantle, so older lithosphere will have greater slab pull. This simple relationship needs some modification, because the oceanic crust, which is bonded to the lithosphere, is less dense (compositional buoyancy) than the underlying mantle, so very young lithosphere will be positively buoyant. An important question is how long the oceanic lithosphere needs to cool before the negative thermal buoyancy exceeds the compositional buoyancy. Oxburgh and Parmentier (1977) have addressed this issue and have estimated the total buoyancy of oceanic lithosphere.

The compositional buoyancy has two components. Mantle upwelling beneath ridges undergoes decompression melting at a depth of about 40 km. This melt migrates to the magma chamber at the ridge axis, where it forms oceanic crust with a normal thickness of 6–7 km and an average density of 2900 km m^{-3}. In addition, ultramafic residues formed by partial melting during the generation of basalt are less dense than undepleted mantle. This zone of depleted mantle has a thickness of about 21 km and an average density ρ_d of 3235 kg m^{-3}, which is less than the normal density of peridotite mantle of 3300 kg m^{-3}. So this adds an additional buoyancy to the oceanic lithosphere.

Oxburgh and Parmentier (1977) have made a quantitative assessment of the overall buoyancy of the lithosphere and define a parameter called the density defect thickness δ as

$$\delta = \int_0^\infty \left[\frac{\rho_m - \rho(z)}{\rho_m} \right] dz \qquad (5.64)$$

where $\rho(z)$ is the density of the lithosphere including the crust, depleted mantle, and cooled lithosphere, and ρ_m is the normal mantle density. The density defect thickness has units of length such that when $\delta < 0$, the entire package is negatively buoyant and can subduct, while when $\delta > 0$, then the package is positively buoyant and will resist subduction. They further divide the density defect thickness into a compositional and a thermal component

$$\delta_{total} = \delta_{comp} + \delta_{thermal}. \qquad (5.65)$$

The compositional part has a contribution from the crust $0.85 \text{ km} = 7 \text{ km} \times (\rho_m - \rho_c)/\rho_m$ and a contribution from the depleted mantle $0.41 \text{ km} = 21 \text{ km} \times (\rho_m - \rho_d)/\rho_m)$ for a total of $\delta_{comp} = 1.3 \text{ km}$. They also point out that areas having thickened crust, such as oceanic plateaus, will have proportionally larger buoyancy, so that a 20 km thick crust will have a buoyancy of $\delta_{comp} = 4.3 \text{ km}$.

The thermal buoyancy depends on the density of the cooled lithosphere given by equation (5.35) $\rho(T) = \rho_m [1 - \alpha(T - T_m)]$. Inserting this into equation (5.64) and simplifying, one finds

$$\delta_{thermal} = -2\alpha (T_m - T_o) \int_0^{\infty} \text{erfc}\left(\frac{z}{2\sqrt{\kappa t}}\right) dz. \qquad (5.66)$$

We performed a similar integration of equation (5.39). The result is

$$\delta_{thermal} = -2\alpha (T_m - T_o) \sqrt{\frac{\kappa t}{\pi}}. \qquad (5.67)$$

The total density defect thickness decreases with increasing age and eventually the thermal buoyancy dominates. An example is shown in Figure 5.8. For a normal crustal thickness of 6 km, the lithosphere is positively buoyant between 0 and 30 Ma and negatively buoyant for greater ages. Lithosphere with thicker crust, such as that associated with oceanic plateaus, remains positively buoyant for a longer time. For example, an 18 km thick oceanic plateau will resist subduction at even the oldest ages found in the ocean basins of 200 Ma. In addition to crustal thickness, the buoyancy depends on the temperature difference across the lithosphere or thermal boundary layer. Lithosphere on the planet Venus will be more buoyant than comparable lithosphere on Earth, because Venus has a higher surface temperature of 455 °C. The calculation of buoyancy versus cooling time for Venus is left as an exercise below.

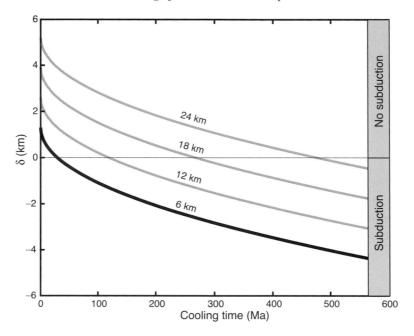

Figure 5.8 Total density defect thickness versus the age (or cooling time) of the lithosphere. Normal oceanic crust is 6 km thick and the lithosphere becomes negatively buoyant after cooling for 30 Ma. For a crustal thickness of 12 km, the time until negative buoyancy is delayed until 120 Ma. Lithosphere having a 24 km thick crust will remain buoyant for 500 Ma. Moreover, the lower lithosphere may be reheated or may delaminate during this long time, so it may never subduct.

5.7 Exercises

Exercise 5.1 (a) What two measurements must be made to determine the conductive heat flow at the bottom of the ocean? (b) Why is it OK to measure heat flow in the upper few meters of sediment on the seafloor, while one needs a borehole hundreds of meters deep to obtain a reliable measure of heat flow on the continents?

Exercise 5.2 Assume that the lithosphere of Venus has evolved to a steady-state temperature profile and there is no heat generated in the lithosphere. Given a current heat flow of 4×10^{-2} W m^{-2}, a surface temperature of 450 °C, a mantle temperature of 1500 °C, and a thermal conductivity of 3.3 W m^{-1} °C^{-1}, calculate the thickness of the lithosphere.

Exercise 5.3 Solve for the temperature T as a function of time t and depth z in a cooling half space. The differential equation for heat diffusion is, and the boundary/initial conditions are:

$$\frac{\partial^2 T}{\partial z^2} = \frac{1}{\kappa} \frac{\partial T}{\partial t}$$

$$
\begin{aligned}
T(0,t) &= T_o \\
T(\infty,t) &= T_m \\
T(z,0) &= T_m.
\end{aligned}
\tag{5.68}
$$

Use the following similarity variable $\eta = \frac{z}{2\sqrt{\kappa t}}$ to reduce the partial differential equation to an ordinary differential equation where κ is the thermal diffusivity, T_o is the surface temperature, and T_m is the initial temperature of the half space.

Exercise 5.4 Derive the following relationship between the rate of increase in seafloor depth with age $\frac{\partial d}{\partial t}$ and the difference between the surface and basal heat flow $(q_s - q_L)$.

$$\frac{\partial d}{\partial t} = \frac{\alpha}{C_p (\rho_m - \rho_w)} (q_s - q_L) \tag{5.69}$$

You will need Fourier's law, energy conservation, and isostasy as follows:

$$q = k \frac{\partial T}{\partial z}$$

$$\frac{\partial T}{\partial t} = \frac{k}{\rho_m C_p} \frac{\partial^2 T}{\partial z^2} \tag{5.70}$$

$$d(t) = \frac{-\alpha \rho_m}{(\rho_m - \rho_w)} \int_o^L T \, dz$$

where:

- L asymptotic lithospheric thickness and also the depth of compensation (m)
- d seafloor depth (m)
- q heat flow (W m^{-2})
- α coefficient of thermal expansion ($^\circ$C^{-1})
- C_p heat capacity (J kg^{-1})
- ρ_w seawater density (kg m^{-3})
- ρ_m mantle density (kg m^{-3})
- k thermal conductivity (W m^{-1} $^\circ$C^{-1})

Exercise 5.5 Calculate the cooling time for lithospheric subduction on Venus for crustal thicknesses of 16 km and 24 km. The surface temperature of Venus is 455 $^\circ$C; use a deep-mantle temperature of 1400 $^\circ$C. Use Earth-like values of thermal expansion coefficient and thermal diffusivity in Table 5.2.

Exercise 5.6 Show that the coefficients of the Fourier series (5.55) are $a_n = \frac{2}{n\pi}$.

Exercise 5.7 Perform the integration in (5.62) to derive the formula for depth versus age given in (5.63).

Exercise 5.8 Seasonal variations in temperature near the surface of a glacier. Derive the formula 9.8 in Chapter 9 of Cuffey and Patterson (2010) and reproduce Figure 9.2. Also discuss the seasonal variations in heat flow.

Exercise 5.9 Temperature evolution of an oceanic fracture zone (Sandwell and Schubert, 1982a). An oceanic fracture zone is the boundary between lithosphere of different ages as shown in Figure 9.15. Consider the profile **A-A'** in Figure 9.15. The temperature far from the FZ near **A** is the deep mantle temperature T_m and far from the FZ on the **A'** side is the error function solution given in equation (5.23). As the FZ ages, there will be vertical diffusion of heat causing additional half-space cooling. Also, there will be lateral heat transport across the FZ from the young (hot) side to old (cold) side. The differential equation, initial condition, and boundary conditions are

$$\frac{\partial^2 T}{\partial x^2} + \frac{\partial^2 T}{\partial z^2} = \frac{1}{\kappa} \frac{\partial T}{\partial t} \tag{5.71}$$

$$
\begin{aligned}
T(x,z,t_o) &= T_m & x < 0 \\
T(x,z,t_o) &= T_o + (T_m - T_o)\operatorname{erf}\left(\frac{z}{\sqrt{\kappa t_o}}\right) & x > 0 \\
T(x,0,t_o) &= T_o \\
T(x,z,t_o) &= T_m
\end{aligned}
\tag{5.72}
$$

where z is depth, x is distance across the FZ from **A** to **A'**, t is the age of the **A** side and t_o is the age of the **A'** side. The solution for the temperature is

$$
\begin{aligned}
T(x,z,t) = T_o + &\frac{(T_m - T_o)}{2}\left[\operatorname{erfc}\frac{x}{2\sqrt{\kappa(t-t_o)}}\operatorname{erf}\frac{z}{2\sqrt{\kappa(t-t_o)}}\right. \\
&\left. + \operatorname{erfc}\frac{-x}{2\sqrt{\kappa(t-t_o)}}\operatorname{erf}\frac{z}{2\sqrt{\kappa t}}\right].
\end{aligned}
\tag{5.73}
$$

(a) Show that this solution satisfies the differential equation, initial condition, and boundary conditions.

(b) Derive equation (5.73). This can be done by direct convolution in the space domain or multiplication in the wavenumber domain. Both are algebraically challenging and cannot be found in the literature.

(c) Write a MATLAB program to create a contour plot or image of the temperature versus x and z for any time $t > t_o$.

Exercise 5.10 Frictional heating during and earthquake. During an earthquake, most of the energy is converted to heat. Calculate the temperature across a fault during and following the earthquake for a variety of fault zone widths. Derive equation (5) in Fialko (2004).

Exercise 5.11 Temperature and heat flow from mantle plume. When the lithosphere passes over a mantle plume, the lower lithosphere is reheated. The motion of the lithosphere will advect the temperature anomaly $T(x, y, z)$ downstream. Also, heat will diffuse vertically through the lithosphere toward the surface, resulting in a heat flow anomaly that is maximum downstream from the source. To simulate this reheating, we set up a problem where a half space is moving at a velocity $\mathbf{v} = (v_x, v_y)$ through a fixed heat source at depth z_o given by $q_o(x, y, z) = q(x, y)\,\delta(z - z_o)$. The differential equation and boundary conditions for the temperature anomaly are

$$\mathbf{v} \cdot \nabla T - \kappa \nabla^2 T = \frac{q_o(x, y, z)}{\rho C_p} \tag{5.74}$$

$$T(x, y, 0) = 0$$

$$\lim_{z \to \infty} T(x, y, z) = 0$$

$$\lim_{|x| \to \infty} T(x, y, z) = 0 \tag{5.75}$$

$$\lim_{|y| \to \infty} T(x, y, z) = 0$$

where κ is the thermal diffusivity, C_p is the heat capacity, and ρ is the density. The solution for an arbitrary heat source at depth z_o is

$$T(\mathbf{k}, z) = \frac{Q(\mathbf{k})}{4\pi \rho C_p \kappa p} \left[e^{-2\pi p |z_o - z|} - e^{-2\pi p |z + z_o|} \right] \tag{5.76}$$

where $\mathbf{k} = (k_x, k_y)$ is the horizontal vector wavenumber and $Q(\mathbf{k})$ is the 2-D Fourier transform of the heat source. The parameter p in the solution is a combination of the wavenumbers and velocity.

$$p^2 = \mathbf{k} \cdot \mathbf{k} - \frac{i}{2\pi \kappa} \mathbf{v} \cdot \mathbf{k} \tag{5.77}$$

(a) Derive equation (5.76). Start by taking the 3-D Fourier transform of equation (5.74) and separate the horizontal wavenumbers from the vertical wavenumber. The result is

$$\left(\mathbf{k} \cdot \mathbf{k} - \frac{i}{2\pi \kappa} \mathbf{v} \cdot \mathbf{k} + k_z^2 \right) T(\mathbf{k}, k_z) = \frac{Q(\mathbf{k})\, e^{-i2\pi k_z z_o}}{4\pi^2 \rho C_p \kappa}. \tag{5.78}$$

Continue with the derivation using definition of p^2 given in equation (5.77). Perform the inverse transform with respect to k_z by integrating around the poles in the complex plane keeping only terms with decaying exponential in z. Finally, use the method of images to satisfy the surface boundary condition.

(b) Compute the vertical heat flow and then the surface heat flow.

(c) Write a MATLAB program to compute the temperature and heat flow at any depth using a Gaussian heat source $q(x, y, z) = A\delta(z - z_o) \exp\left(-\frac{x^2 + y^2}{2\sigma^2}\right)$ where σ is the half-radius of the source. The computational approach is to generate a 2-D array representing the Gaussian heat source. Put the center of the source somewhere in the

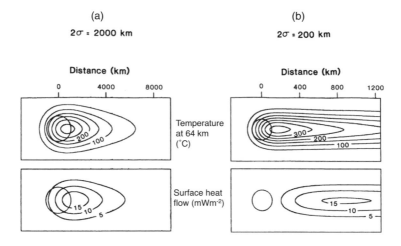

Figure 5.9 The response of the lithosphere moving at a velocity of 40 mm/yr over a Gaussian heat source with half widths (circles) of (a) 2000 km and (b) 200 km. The maximum temperature is approximately at the downstream edge of the source while the surface heat flow is displaced significantly downstream (Sandwell, 1982).

middle of the 2-D array. Take its 2-D Fourier transform. Generate the wavenumbers and the solution in equation (5.76). Multiply the source and solution and perform the inverse 2-D transform. Note the equation for the temperature (5.76) is singular when both horizontal wavenumbers are zero (i.e., $p = 0$). Set this term to zero prior to computing the 2-D inverse Fourier transform. To simulate the Hawaiian plume, use an x-velocity of 40 mm/yr and a 2σ of 1000 km and place the source at a depth of 60 km. Make the width of the array 4000 km and the length of the array in the x-direction at least 20,000 km. This great length is needed to avoid the Fourier edge effect caused by heat flowing to great depth. Make images or contour maps of the temperature at a depth of 50 km and the surface heat flow making sure the pixels are square so the results are not distorted. Explore the parameter space of plume strength, A, plume radius σ, plume depth z_o, and plate speed v_x. Note that if a finite thickness plate is used both the wrap-around and p-singularity will be gone (Sandwell, 1982). An example of the temperature and surface heat flow for a thinned lithosphere is shown in Figure 5.9.

6

A Brief Review of Elasticity

This is a very brief review of the elasticity theory needed to understand the principles of stress, strain, and flexure in *Geodynamics* (Turcotte and Schubert, 2014). This review assumes that you have already taken a course in continuum mechanics. One difference from *Geodynamics* is that we follow the sign convention used by seismologists and engineers, where extensional strain and stress are positive.

6.1 Stress

Stress is a force acting on an area and is measured in newtons per meter squared ($N\,m^{-2}$), which corresponds to a pascal unit (Pa). Figure 6.1 shows a cube of solid material. Each face of the cube has three components of stress, so there are nine possible components of the stress tensor.

We will consider only the symmetric part of the stress tensor, so only six of these components are independent. The antisymmetric part of the tensor represents a torque. In Cartesian coordinates, the stress tensor is given by

$$
\sigma_{ij} = \begin{bmatrix} \sigma_{xx} & \sigma_{xy} & \sigma_{xz} \\ \sigma_{xy} & \sigma_{yy} & \sigma_{yz} \\ \sigma_{xz} & \sigma_{yz} & \sigma_{zz} \end{bmatrix} \tag{6.1}
$$

where index notation is the shorthand for dealing with tensors and vectors; a variable with a single subscript is a vector $\vec{a} = a_i$, a variable with two subscripts is a tensor $\sigma = \sigma_{ij}$, and a repeated index indicates summation over the spatial coordinates. For example, the pressure is given by $P = -\sigma_{ii}/3$. In addition, a comma preceding a subscript means differentiation with respect to that variable $\nabla \vec{a} = a_{i,j}$ or, for example, $a_{x,y} = \frac{\partial a_x}{\partial y}$.

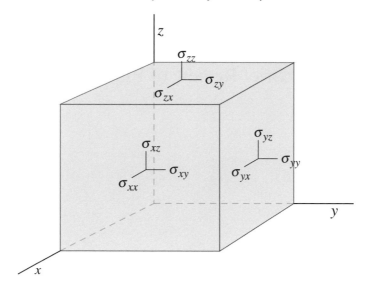

Figure 6.1

6.2 Strain

Strain is change in length over the original length, so it is a dimensionless variable; we will assume strains are small ($\ll 10^{-3}$). Let the displacement vector field inside of a solid body be given by

$$\vec{u} = u_i = \begin{bmatrix} u_x & u_y & u_z \end{bmatrix} \tag{6.2}$$

The gradient of this vector is a tensor: $\nabla \vec{u} = u_{i,j}$. This tensor is commonly decomposed into a symmetric tensor (strain) and an antisymmetric tensor (rotation):

$$u_{i,j} = \frac{1}{2}\left(\frac{\partial u_i}{\partial x_j} + \frac{\partial u_j}{\partial x_i} \right) + \frac{1}{2}\left(\frac{\partial u_i}{\partial x_j} - \frac{\partial u_j}{\partial x_i} \right) \tag{6.3}$$

We will not consider the rotation tensor further, but the strain tensor is given by

$$\varepsilon_{ij} = \tfrac{1}{2}\left(u_{i,j} + u_{j,i} \right) \tag{6.4}$$

6.3 Stress versus Strain

If one assumes the material has an isotropic and linear response, then the relationship between stress and strain is given by

$$\sigma_{ij} = \lambda\, \delta_{ij}\, \varepsilon_{kk} + 2\mu\, \varepsilon_{ij} \tag{6.5}$$

where δ_{ij} is equal to 0, except when $i = j$; then it is equal to 1. The Lamé constants λ and μ define the elastic properties. The shear modulus μ (or G in the engineering literature) relates the shear stress to shear strain on a component-by-component basis:

$$\sigma_{xy} = 2\mu\varepsilon_{xy} = \mu\left(\frac{\partial u_x}{\partial y} + \frac{\partial u_y}{\partial x}\right) \tag{6.6}$$

6.4 Principal Stress and Invariants

This general relation between stress and strain tensors is rather involved. It is difficult to invert this relationship to develop a relationship between strain and stress. One means of simplifying this relationship is to find a coordinate system rotation **R** that will cause the stress and strain tensors to be diagonal. Real symmetric matrices have real eigenvalues, orthogonal eigenvectors, and can be diagonalized. This implies that there always exists some principal coordinate system where the shear stresses are zero on planes orthogonal to the coordinate axes, and the normal stresses act along the principal axes directions. The eigenvectors form the rotation matrix **R** and the eigenvalues form the principal stress tensor

$$\vec{\sigma}_p = \begin{bmatrix} \sigma_1 & 0 & 0 \\ 0 & \sigma_2 & 0 \\ 0 & 0 & \sigma_3 \end{bmatrix} = \mathbf{R}\sigma\mathbf{R}^t \tag{6.7}$$

where they are ordered from largest to smallest

$$\sigma_1 \geq \sigma_2 \geq \sigma_3. \tag{6.8}$$

To establish this rotation, consider a block of material having a uniform stress state given by

$$\begin{bmatrix} \sigma_{xx} & \sigma_{xy} & \sigma_{xz} \\ \sigma_{xy} & \sigma_{yy} & \sigma_{yz} \\ \sigma_{xz} & \sigma_{yz} & \sigma_{zz} \end{bmatrix}. \tag{6.9}$$

We seek a planar surface within the block where there is only a normal traction γ and no shear tractions. In other words, given the vector **n** that is normal to that surface we seek **n** such that

$$\begin{bmatrix} \sigma_{xx} & \sigma_{xy} & \sigma_{xz} \\ \sigma_{xy} & \sigma_{yy} & \sigma_{yz} \\ \sigma_{xz} & \sigma_{yz} & \sigma_{zz} \end{bmatrix} \begin{bmatrix} n_x \\ n_y \\ n_z \end{bmatrix} = \begin{bmatrix} \gamma n_x \\ \gamma n_y \\ \gamma n_z \end{bmatrix}. \tag{6.10}$$

We can rewrite this equation as

$$
\begin{bmatrix}
\sigma_{xx} - \gamma & \sigma_{xy} & \sigma_{xz} \\
\sigma_{xy} & \sigma_{yy} - \gamma & \sigma_{yz} \\
\sigma_{xz} & \sigma_{yz} & \sigma_{zz} - \gamma
\end{bmatrix}
\begin{bmatrix}
n_x \\
n_y \\
n_z
\end{bmatrix}
= 0. \tag{6.11}
$$

This traction γ and matching normal vector **n** will be one of the three principal stresses and principal stress directions, respectively.

There are three properties (*invariants*) of the stress tensor that do not change under coordinate rotation. The invariants are found by first developing the characteristic equation from the determinant of the following equation

$$
\begin{vmatrix}
\sigma_{xx} - \gamma & \sigma_{xy} & \sigma_{xz} \\
\sigma_{xy} & \sigma_{yy} - \gamma & \sigma_{yz} \\
\sigma_{xz} & \sigma_{yz} & \sigma_{zz} - \gamma
\end{vmatrix}
= 0 \tag{6.12}
$$

which becomes

$$
\gamma^3 - I\gamma^2 + II\gamma - III = 0 \tag{6.13}
$$

where the stress invariants are

$$
I = \sigma_{ii}
$$
$$
II = \frac{1}{2}\left(\sigma_{ii}\sigma_{jj} - \sigma_{ij}\sigma_{ij}\right) = \sigma_{xx}\sigma_{yy} + \sigma_{yy}\sigma_{zz} + \sigma_{xx}\sigma_{zz} - \sigma_{xy}^2 - \sigma_{yz}^2 - \sigma_{xz}^2 \tag{6.14}
$$
$$
III = \left|\sigma_{ij}\right|
$$

the trace I, the sum of minors II, and the determinant of the stress tensor III. The first invariant is related to the mean normal stress or pressure $P = -I/3$. The second invariant is related to shear stress and thus is commonly used as the Von Mises failure criteria. We will not consider the third invariant further.

> **Exercise 6.1** Use symbolic algebra in MATLAB to take the determinant of the characteristic equation (6.12). Identify the first and second invariants in the third-order polynomial and check that they match the invariants in equation (6.14).

The principal stress system is important in geophysics and geology. Owing to the presence of the free surface, the stress field close to the Earth's surface is expected to have one principal stress vertical and hence two horizontal principal stresses. Also, in the Earth, we sometimes subtract the pressure from the stress tensor.

In this case, it is called *deviatoric stress*. In the principal stress system, the pressure and maximum shear stress are given by

$$P = -\frac{1}{3}(\sigma_1 + \sigma_2 + \sigma_3)$$

$$\tau = \frac{1}{2}(\sigma_1 - \sigma_3). \tag{6.15}$$

6.5 Principal Stress and Strain

The stress versus strain relation is far simpler in the principal coordinate system

$$\begin{bmatrix} \sigma_1 \\ \sigma_2 \\ \sigma_3 \end{bmatrix} = \begin{bmatrix} \lambda + 2\mu & \lambda & \lambda \\ \lambda & \lambda + 2\mu & \lambda \\ \lambda & \lambda & \lambda + 2\mu \end{bmatrix} \begin{bmatrix} \varepsilon_1 \\ \varepsilon_2 \\ \varepsilon_3 \end{bmatrix} \tag{6.16}$$

where ε_1, ε_2, and ε_3 are the principal strains. Next, we can use this relationship to develop three important parameters: Poisson's ratio ν, Young's modulus E, and bulk modulus K.

First, consider the case of uniaxial stress where $\sigma_2 = \sigma_3 = 0$. This represents application of an end load to an elastic beam fastened to a wall. The second equation for σ_2 is

$$0 = \lambda\varepsilon_1 + (\lambda + 2\mu)\varepsilon_2 + \lambda\varepsilon_3. \tag{6.17}$$

Because of symmetry, we know $\varepsilon_2 = \varepsilon_3$, so we arrive at a relationship between ε_2 and ε_1:

$$\varepsilon_2 = \frac{-\lambda}{2(\lambda + \mu)}\varepsilon_1 = -\nu\varepsilon_1 \tag{6.18}$$

where ν is Poisson's ratio. Next, we can use this relationship between strains in the first equation to provide a relationship between σ_1 and ε_1.

$$\sigma_1 = (\lambda + 2\mu)\varepsilon_1 + \frac{-\lambda^2}{\lambda + \mu}\varepsilon_1$$

$$\sigma_1 = \frac{(\lambda + 2\mu)(\lambda + \mu) - \lambda^2}{\lambda + \mu}\varepsilon_1$$

$$\sigma_1 = \frac{\mu(3\lambda + 2\mu)}{\lambda + \mu}\varepsilon_1 \tag{6.19}$$

$$\sigma_1 = E\varepsilon_1$$

where E is Young's modulus.

Next we consider the case of uniform pressure. In this case, the change in pressure $\Delta P = -\left(\sigma_1 + \sigma_2 + \sigma_3\right)/3$ is related to a change in volume $\Delta V = \left(\varepsilon_1 + \varepsilon_2 + \varepsilon_3\right)$. Using the stress–strain relation, we find

$$\Delta P = -\left(\lambda + \tfrac{2}{3}\mu\right)\Delta V$$

$$\Delta P = -K\Delta V \tag{6.20}$$

where K is the bulk modulus. One can invert this stress versus strain relationship (equation (6.16)) to obtain a strain versus stress relationship. We'll also assume that the principal coordinates are aligned with the x, y, and z axes.

$$\begin{bmatrix} \varepsilon_1 \\ \varepsilon_2 \\ \varepsilon_3 \end{bmatrix} = \frac{1}{E}\begin{bmatrix} 1 & -\nu & -\nu \\ -\nu & 1 & -\nu \\ -\nu & -\nu & 1 \end{bmatrix}\begin{bmatrix} \sigma_1 \\ \sigma_2 \\ \sigma_3 \end{bmatrix} \tag{6.21}$$

Exercise 6.2 Use symbolic algebra in MATLAB to invert the stress versus strain relationship (equation (6.16)) to obtain the relationship between strain and stress (equation (6.21)). Show that the product of these two matrices is the identity matrix.

Now we have arrived at equations (3.31), (3.32), and (3.33) in *Geodynamics* (Turcotte and Schubert, 2014). Before moving onto the flexure problem, we consider the case of a thin elastic plate. "Thin plate" means that there are no variations in the vertical displacement field as a function of depth in the plate, so we can make the approximation $\sigma_{zz} = \sigma_3 = 0$. Under this approximation, we have the following:

$$\varepsilon_{xx} = \tfrac{1}{E}\left(\sigma_{xx} - \nu\sigma_{yy}\right)$$

$$\varepsilon_{yy} = \tfrac{1}{E}\left(\sigma_{yy} - \nu\sigma_{xx}\right) \tag{6.22}$$

$$\varepsilon_{zz} = \tfrac{-\nu}{E}\left(\sigma_{xx} + \sigma_{yy}\right)$$

These equations are the starting point for the development of the relationship between bending moment and curvature provided in *Geodynamics* (Turcotte and Schubert, 2014, Section 3.9).

6.6 Exercises

Exercise 6.3 Use the thin plate equations (equation (6.22)) to develop a linear relationship between moment and curvature. What are the important parameters that control the *flexural rigidity*? It will be helpful to study T&S (Turcotte and Schubert, 2014, Section 3.9).

7

Crustal Structure, Isostasy, Swell Push Force, and Rheology

7.1 Introduction

This chapter covers four topics. First, the basic structure of the oceanic and continental crust is provided. The emphasis is on layer thickness and densities, and there is little discussion of composition. The second and third topics are the vertical and horizontal force balances due to variation in crustal thickness. The vertical force balance, *isostasy*, provides a remarkably accurate description of variations in crustal thickness based on a knowledge of the topography. The horizontal force balance provides a lower bound on the force needed to maintain topographic variations on the Earth. The basic question is: "What keeps mountain ranges from spreading laterally under their own weight?"

The fourth topic is the rheology of the lithosphere (Brace and Kohlstedt, 1980). How does the lithosphere strain in response to applied deviatoric stress? The uppermost part of the lithosphere is cold, so frictional sliding along optimally oriented, pre-existing faults governs the strength. At greater depth, the rocks can yield by non-linear flow mechanisms. The overall strength-versus-depth profile is called the *yield strength envelope* (YSE). The integrated yield strength transmits the global plate tectonic stress. Moreover, the driving forces of plate tectonics cannot exceed the integrated lithospheric strength. This provides an important constraint on the geodynamics of oceans and continents.

The yield-strength-envelope formulation will also be used in the chapter on flexure. It provides an explanation for the increase in the thickness of the elastic layer as the lithosphere ages and cools. In addition, it is used to understand the depth of oceanic trenches. The first moment of the yield strength versus depth provides an upper bound on the magnitude of the bending moment that the lithosphere can maintain. This model strength estimate can be checked by measuring the bending moment of the trench/outer rise topography. There is remarkably good agreement, assuming mantle dynamics does not play a large role in the support of subduction zone topography. Throughout these two chapters, the deviatoric stress levels are typically 100–300 MPa, since this is the level of stress needed to maintain the topographic features on the Earth. In the following chapters on earthquakes, the magnitudes of stresses are typically less than 10 MPa. Part of this order-of-magnitude stress reduction is due to the fact that many earthquakes are shallow. However, there is still a major unresolved issue of why the crust appears weak in regard to earthquakes and strong in regard to topography and flexure. Since this is a major unresolved issue, it is a good topic for research.

7.2 Oceanic Crustal Structure

The basic structure of the oceanic crust has been established through seismic refraction and reflection experiments, seafloor dredging/drilling, gravity/magnetics modeling, and studies of ophiolites (Ryan, 1994). Figure 7.1 from Kent et al.

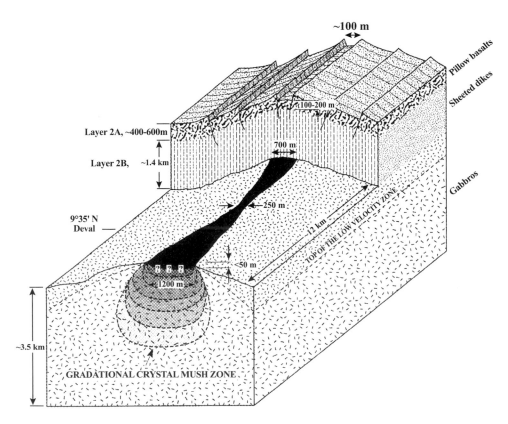

Figure 7.1 Model of crustal structure derived from reflection and refraction seismology (Kent et al., 1993).

Table 7.1. *Structure of oceanic crust. The total crustal thickness is 5–7 km (Chen, 1992) and older crust is slightly thicker (Van Avendonk et al., 2017).*

Oceanic Crust	Density (kg m^{-3})	Compressional Velocity (km s^{-1})	Thickness (km)
seawater	1025	1.5	2–6
sediment	>2300	> 1.65	0–20
basalt	2700	5.0	2
gabbro	2800	6.8	4
mantle peridotite	3325	8.15	—

(1993) illustrates the current model of crustal generation at a fast-spreading ridge. Partial melt that forms by pressure release in the uppermost mantle (~40 km depth) percolates to a depth of about 2000 m beneath the ridge, where it accumulates to form a thin magma lens. Beneath the lens is a mush zone which develops into a 3500 m thick gabbro layer by some complicated ductile flow. Above the lens, sheeted dikes (~1400 m thick) are injected into the widening crack at the ridge axis. Part of this volcanism is extruded into the seafloor as pillow basalts. The pillow basalts and sheeted dikes cool rapidly as cool seawater percolates to a depth of at least 2000 m. This process forms the basic crustal layers seen by reflection and refraction seismology methods (Table 7.1).

Some key points about the oceanic crust:

- The Mohorovicic Discontinunity (Moho) is defined as the seismic velocity jump from 6.8 km s^{-1} to greater than 8 km s^{-1}.
- The Moho corresponds to a change in composition and density. For thick continental crust, it also corresponds to a change in strength.
- The oceanic crustal thickness is remarkably uniform throughout the ocean basins. While the average crustal thickness does not depend on spreading rate, the local variations in thickness are greater for the slow-spreading crust (<70 mm/yr full rate). There are thickness variations along a ridge segment bounded by transform faults; the crust is generally thinner near the end of a spreading segment and thicker toward the center of the spreading segment.

7.3 Continental Crustal Structure

The basic structure of the continental crust has also been established through seismic refraction and reflection experiments, sampling and drilling, gravity/magnetics

Table 7.2. *Structure of continental crust. Total crustal thickness is 34 km for zero elevation.*

Continental Crust	Density (kg m^{-3})	Compressional Velocity (km s^{-1})	Thickness (km)
sediment	>2300	> 5.0	0–20
upper crust	2800	6.3	15
lower crust	2900	6.8	20
mantle peridotite	3320	8.1	—

modeling, and studies of exposed crustal sequences. Of course, the continental crust is highly variable in thickness, velocity, density, and composition. Table 7.2 represents an average crustal model for elevations close to sea level.

7.4 Vertical Force Balance: *Isostasy*

As discussed in Chapter 1, one of the most important and defining features of the Earth is the bimodal histogram of topography (Figure 1.2). The tallest peak in the histogram represents continental crust having elevations close to sea level. A second broader peak represents the oceanic crust having elevations between about -6000 m and -3000 m with a median depth of -4093 m. This bimodal histogram can be largely explained by simple Airy isostasy, where the lithostatic pressure at the base of the lithosphere is constant over the Earth.

A diagram of the Airy compensation model is shown in Figure 7.2 (Schubert and Sandwell, 1989). A uniform density crust is divided into five layers, which float on the higher density mantle. A major assumption of this model is that the mantle beneath the crustal plateau has no lateral density variations. To understand how the thickness of each layer is calculated, first consider a plateau on the seafloor where all of the topography lies below sea level. By definition, the thicknesses of layers 1 and 5 are zero. The topography of the plateau above the normal seafloor depth (h_2) is isostatically compensated by a crustal root with a thickness of h_4. Isostatic balance means the pressure at the base of the crustal root is the same as the pressure at the same depth beneath normal oceanic crust. The balance is

$$g \left(h_2\, \rho_c + h_3\, \rho_c + h_4\, \rho_c \right) = g \left(h_2\, \rho_w + h_3\, \rho_c + h_4\, \rho_m \right). \tag{7.1}$$

This can be simplified to an equation where the root thickness is related to the elevation:

$$h_4 = \frac{(\rho_c - \rho_w)}{(\rho_m - \rho_c)} h_2 \tag{7.2}$$

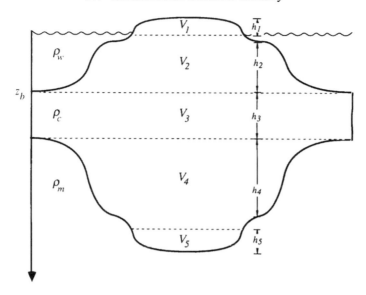

Figure 7.2 Airy compensation model used to determine crustal thickness and crustal volume. The uniform density crust ($\rho_c = 2800$ kg m^{-3}) is divided into five layers which float on the higher density mantle ($\rho_m = 3200$ kg m^{-3}). The seawater has a density ρ_w, of 1025 kg m^{-3}. Layer 1 lies above sea level, layer 2 lies between sea level and the normal seafloor depth (base depth) z_b, layer 3 corresponds to the thickness of normal oceanic crust, layer 4 is the compensating root for layer 2, and layer 5 is the compensating root for layer 1.

The densities determine the ratio of root thickness to elevation; for typical values provided in Figure 7.2, this ratio is 4.4. The overall crustal thickness is the sum of the three layers

$$h_t = h_3 + h_2 \left[1 + \frac{(\rho_c - \rho_w)}{(\rho_m - \rho_c)} \right]. \tag{7.3}$$

Exercise 7.1

(a) Develop a formula for the total crustal thickness when the top of the plateau is at sea level. Use the densities provided in Figure 7.2, a normal oceanic crustal thickness of 6.5 km, and a normal seafloor depth of 4.1 km.
(b) Calculate the thickness of this plateau at sea level.
(c) How does this compare with a typical value for continental crustal thickness at sea level?

This simple Airy isostasy model provides a first-order explanation for spatial variations in crustal thickness observed using seismic refraction measurements in both the oceans (Figure 7.3) and continents (Figure 7.4). For a continent elevation

Plateaus

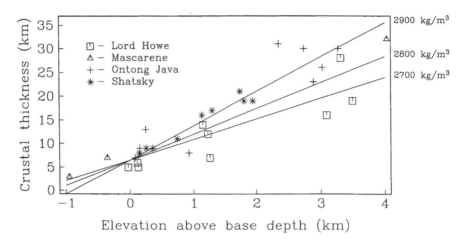

Figure 7.3 Seismic refraction measurements of crustal thickness versus elevation above base depth for two continental submarine plateaus (Lord Howe Rise and Mascarene Plateau) and two oceanic plateaus (Ontong-Java and Shatsky Rise) (Schubert and Sandwell, 1989). Predictions of the Airy compensation model are shown for crustal densities ranging from 2700 to 2900 kg m^{-3} and a mantle density of 3200 kg m^{-3}.

Continents

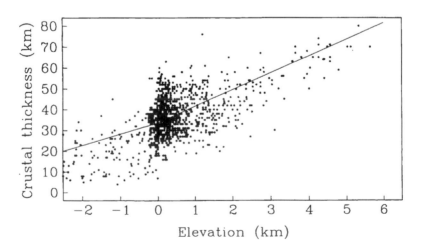

Figure 7.4 Seismic refraction measurements of crustal thickness versus elevation (Schubert and Sandwell, 1989). Measurements were selected from continental areas. Antarctica and Greenland data were not used. The best-fitting Airy compensation model (solid curve) has a zero-elevation crustal thickness of 34 km. The RMS scatter about the model is 9.09 km.

of 5 km, such as the Tibetan plateau, this simple model predicts the crust is 70 km thick—which is in fair agreement with the seismic measurements.

The scatter in the seismic thickness with respect to the estimates from Airy isostasy (Figure 7.4) can be caused by three factors. First, there may be significant lateral variations in crustal density (i.e., Pratt compensation). Second, there may be density variations in the mantle lithosphere caused by thermal or compositional variations. The most prominent example is the elevation of the seafloor spreading ridges (−2500 m) above the average basin depth (−4100 m) caused by thermal isostasy of the cooling oceanic lithosphere (Chapter 5). Finally, the assumption of constant pressure at the base of the lithosphere could be incorrect, because of pressure variations due to mantle flow (i.e., dynamic topography). While these processes are important, Airy isostasy provides a first-order explanation for the bimodal elevation of the Earth.

7.5 Horizontal Force Balance: *Swell Push Force*

One important question that arises is what the magnitude of the stress in the lithosphere needs to be to maintain large-scale plateaus and roots (Flesch et al., 2001). This can be considered as a minimum deviatoric stress, and it places constraints on the long-term strength of the crust—especially the lower crust. Consider isostatically compensated topography as shown in Figure 7.5.

While this diagram is related to a specific Airy-type compensation mechanism, the integral relation, presented next, is quite general. To calculate the total outward force F_s due to this isostatically compensated plateau, we integrate the difference

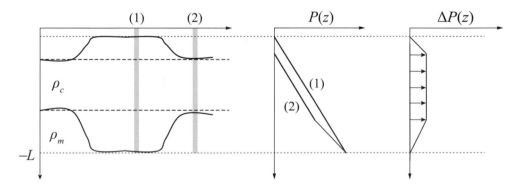

Figure 7.5 (left) Airy-compensated plateau. Pressure at depth −L is constant. (center) Pressure versus depth for columns (1) and (2). (right) Pressure difference versus depth (1) minus (2).

in pressure between column (1) and column (2) over depth, to the depth of compensation $-L$, where the pressure difference is zero:

$$F_s = \int_{-L}^{0} \Delta P(z)\, dz \qquad (7.4)$$

Integrate by parts:

$$F_s = \Delta P(z)\, z \big|_{-L}^{0} - \int_{-L}^{0} \frac{\partial \Delta P(z)}{\partial z} z\, dz \qquad (7.5)$$

Note the first term on the right is zero because of isostasy. The second term can be written in terms of the density by noting that the vertical gradient in the pressure difference is

$$\frac{\partial \Delta P(z)}{\partial z} = -g \Delta \rho(z). \qquad (7.6)$$

The result is

$$F_s = g \int_{-L}^{0} \Delta \rho(z)\, z\, dz. \qquad (7.7)$$

The swell push force depends on two factors: (1) the magnitude of the depth-integrated surface density contrast, which is equal and opposite to the magnitude of the depth-integrated compensation, and (2) the distance between the surface and compensating density. A larger distance between the topography and its compensation increases the swell push force.

Swell push force can be computed for a variety of isostatic configurations; we consider three here. The first case is the calculation of the average crustal stress needed to maintain the elevation of Tibet with respect to the elevation of India. This is left as an exercise (Exercise 7.5); however, we provide the answer of 98 MPa to entice you to do the problem. The important conclusion is that the lower crust must have a strength greater than 98 MPa, or the high elevation of Tibet would rapidly collapse by lateral spreading. Of course, the ongoing collision of India with Asia helps to provide some dynamic support. Nevertheless, there are mountainous areas not having dynamic support that can maintain their high elevation, so the lower crust must be quite strong.

The second case is the calculation of the minimum ice strength needed to maintain the configuration of a floating ice shelf. This is also left as an exercise

(Exercise 7.6). A prominent example of ice sheet failure is the collapse of the Larsen B ice shelf in 2002. Prior to collapse, the ice shelf was the size of Rhode Island. The entire collapse occurred in a couple of months, leaving behind thousands of large ice fragments. A team of collaborating investigators have developed a theory of why the ice disintegrates. The theory is based on the presence of ponded melt-water on the ice shelf surface in late summer, as the climate has warmed in the area. Meltwater acts to enhance fracturing of the shelf, thus reducing its cohesive strength.

The third case is the so-called *ridge push* force. The name is a bit confusing, because the force is zero at the ridge axis and increases with the age of the lithosphere. It is really a gravitational sliding force and is commonly termed *gravitational potential energy* or GPE. For half-space cooling, the seafloor depth increases as the square root of age. The depth of compensation also increases as the square root of age, so the ridge push force increases linearly with age. To calculate the ridge push force, we first construct a density structure following the development in Section 5.4 on thermal subsidence. Figure 7.6 shows this density structure where the temperature T from the cooling half space model (equation 5.23) is

$$T(z,t) = (T_m - T_o)\operatorname{erf}\left(\frac{z}{2\sqrt{\kappa t}}\right) + T_o. \tag{7.8}$$

The depth versus age is provided in equation 5.41 and is

$$d(t) = \frac{2\rho_m \alpha (T_m - T_o)}{(\rho_m - \rho_w)}\left(\frac{\kappa t}{\pi}\right)^{1/2}. \tag{7.9}$$

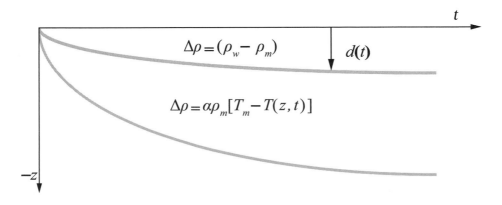

Figure 7.6

The swell push force is

$$F = \int_{-d(t)}^{0} g\,(\rho_w - \rho_m)\,z\,\mathrm{d}z + \int_{-\infty}^{-d(t)} g\alpha\,\rho_m\,[T_m - T\,(z - d,t)]\,(z - d)\,\mathrm{d}z. \quad (7.10)$$

The first integral becomes

$$(1) = g\,(\rho_w - \rho_m)\,\frac{z^2}{2}\Big|_{-d(t)}^{0} = \frac{g}{2}\,(\rho_m - \rho_w)\,d^2\,(t). \quad (7.11)$$

For the second integral, let $z' = d\,(t) - z$, so the integral becomes

$$(2) = g\alpha\,(T_m - T_o)\int_{0}^{\infty} \mathrm{erfc}\left(\frac{z'}{2\sqrt{\kappa t}}\right)z'\,\mathrm{d}z'. \quad (7.12)$$

Now let $\eta = \frac{z'}{2\sqrt{\kappa t}}$, so the integral becomes

$$(2) = g\alpha\,(T_m - T_o)\,4\kappa t\int_{0}^{\infty} \mathrm{erfc}\,(\eta)\,\eta\,\mathrm{d}\eta. \quad (7.13)$$

Now integrate by parts:

$$\int_{0}^{\infty} \mathrm{erfc}\,(\eta)\,\eta\,\mathrm{d}\eta = \mathrm{erfc}\,(\eta)\,\frac{\eta^2}{2}\Big|_{0}^{\infty} + \frac{1}{\sqrt{\pi}}\int_{0}^{\infty} \eta^2 e^{-\eta^2}\,\mathrm{d}\eta \quad (7.14)$$

The first term on the right side is zero, while the second term is $\frac{1}{4}$, so the second integral is simply

$$(2) = g\alpha\,\rho_m\,(T_m - T_o)\,\kappa t. \quad (7.15)$$

The final result is

$$F = g\alpha\rho_m\,(T_m - T_o)\,\kappa\left[1 + \frac{2\alpha\rho_m\,(T_m - T_o)}{\pi\,(\rho_m - \rho_w)}\right]t. \quad (7.16)$$

Now let's put in some numbers. The ridge push force at an age of 100 Ma is 3.2×10^{12} N m^{-1}. The lithosphere is approximately 100 km thick at that age, so the average ridge push stress is 32 MPa.

7.6 Rheology of the Lithosphere

As discussed in the introduction to this chapter, the finite strength of the lithosphere has a dominant effect on plate boundary deformation, intraplate deformation, and lithospheric flexure. For example, the distribution of shallow earthquakes (Figure 1.4) is very different between the oceans and continents. In areas of thin oceanic crust (6–7 km), earthquakes primarily occur in narrow zones following the spreading ridges, transform faults, and subduction zones. In contrast, areas of thicker continental crust (∼35 km) have more diffuse patterns of earthquakes occurring within the broader continental deformation zones. Moreover, we will see in the next two chapters that lithospheric flexure produced by relatively small loads from volcanic growth results in primarily elastic strains in the flexed plate. However, the large amplitude plate bending at ocean trenches results in strains that are well beyond the elastic limits resulting in large normal faults on the outer trench wall and very high plate curvatures suggestive of a thinner than expected elastic lithosphere. These processes are well explained by considering the finite strength of the lithosphere that depends on lithostatic pressure, temperature, rock type, and strain rate.

7.6.1 Overview

A schematic diagram of these processes is shown in Figure 7.7. Pressure increases with depth as $\rho g z$ (Figure 7.7a). According to Byerlee (1978) the maximum shear stress that can be maintained in the upper lithosphere is controlled by frictional sliding on optimally oriented, pre-existing faults. Experimental results show sliding friction f is largely independent of rock type so the maximum stress increases linearly with depth (Figure 7.7b) due to the linear increase in normal stress with depth. As discussed in previous chapters, temperature also increases with depth (Figure 7.7c) depending on factors such as plate age, asymptotic plate thickness, and radiogenic heat sources mostly in the continental crust. As the temperature increases, the lower lithosphere will undergo ductile flow. Using typical geological strain rates of 10^{-17} s^{-1} to 10^{-14} s^{-1}, laboratory experiments for a variety of rock types can be used to constrain the maximum shear stress versus depth in the ductile deformation zone (Figure 7.7d). In response to applied end load or bending moment, the weaker of the two mechanisms will dominate (Figure 7.7e). The maximum frictional sliding strength depends on whether the horizontal external stress is extensional (+) or compressional (−). The overall plot of maximum stress versus in extension and compression is called a *yield strength envelope* (YSE). The transition from frictional sliding (brittle) to ductile flow occurs in a poorly defined brittle-ductile transition zone (Kohlstedt et al., 1995). There are many excellent

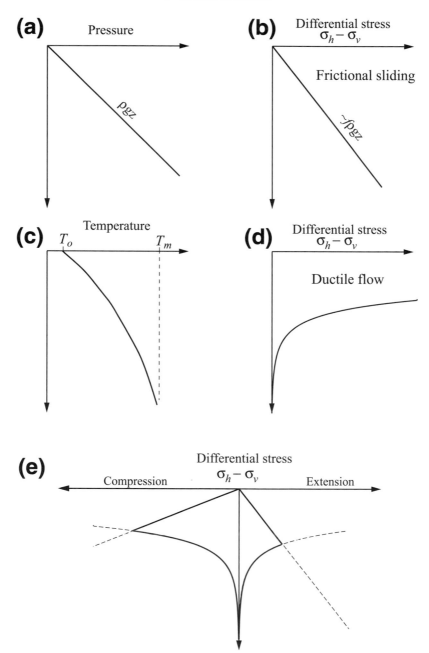

Figure 7.7 Schematic diagram of the components of the yield strength envelope model. (a) Pressure increases with depth. (b) Stress needed to cause frictional sliding on a fault is approximately the coefficient of friction times the lithostatic pressure. (c) Temperature increase with depth. (d) Stress needed to activate ductile flow decreases with increasing temperature. (e) Yield strength versus depth for extension and compression where one principal stress is vertical and a second principal stress is horizontal.

papers on the yield strength envelope. This section follows from four studies – Byerlee (1978), Brace and Kohlstedt (1980), Kohlstedt et al. (1995), and Jaeger et al. (2009). We also repeat several of the analyses and figures provided in the book *Isostasy and Flexure of the Lithosphere* by Watts (2001).

7.6.2 Frictional Sliding on Faults

At depths greater than a few kilometers in the Earth, the cohesive strength of rock is relatively low compared with the overburden pressure ρgz. Over geological timescale, the crust and lithosphere undergo significant deformation creating faults and fractures on many surfaces and in many directions. A classic paper by Byerlee (1978) proposes that fault motion (seismic or aseismic) always occurs on pre-existing faults. Therefore, understanding frictional sliding between rocks at high lithostatic pressure is important for understanding the strength of the upper brittle lithosphere. In that study Byerlee compiled results from rock friction experiments in the intermediate pressure range of 30–2000 MPa corresponding to depths in the Earth of 1–70 km (Figure 7.8). The lithostatic pressure applies a normal stress σ_n perpendicular to the fault plane. The experiments then measure the amount of shear stress τ needed to cause frictional slip. The ratio of shear stress to normal stress is approximately the coefficient of friction. The experimental data show that the relationship is largely independent of rock type and has the functional form

$$\tau = \begin{cases} 0.85\sigma_n & \sigma_n < 200 \text{ MPa} \\ 50 \text{ MPa} + 0.6\sigma_n & \sigma_n > 200 \text{ MPa}. \end{cases} \tag{7.17}$$

We will refer to this as Byerlee's law. Note that in these experiments the coefficient of friction is ~ 0.6 to 0.85. For a vertical strike-slip fault in the crust of density ρ_c, the normal stress is equal to the lithostatic pressure $\rho_c gz$. However, if the fault is filled with water at hydrostatic pressure, then the outward pressure of the water reduces the normal stress such that $\sigma_n = (\rho_c - \rho_w)\,gz$. For example, the shear stress needed to activate a strike-slip fault at 10 km depth in the crust under hydrostatic conditions is quite large ~ 100 MPa. Later in Chapter 11 we will discuss the implications of this stress for heating of the fault zone over many earthquake cycles.

To develop the frictional sliding part of the yield strength envelope consider a layer of rock where one of the principal stresses σ_3 is vertical and the other is horizontal and extensional $\sigma_1 > \sigma_3$ as shown in Figure 7.9. We assume that the layer is extensively fractured, so frictional sliding can occur on faults in any orientation $\theta + \pi/2$ with respect to the x-axis. Given this configuration, we ask the questions what stress difference is needed to induce sliding on an optimally oriented fault and what is that fault orientation?

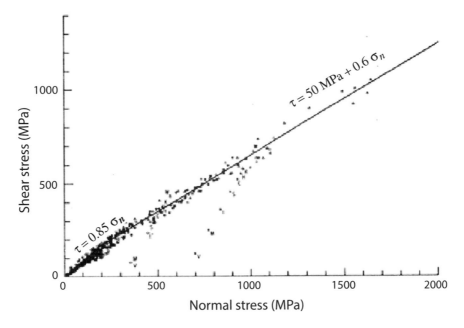

Figure 7.8 Frictional strength for a wide variety of rocks plotted as a function of normal stress (modified from Byerlee (1978)).

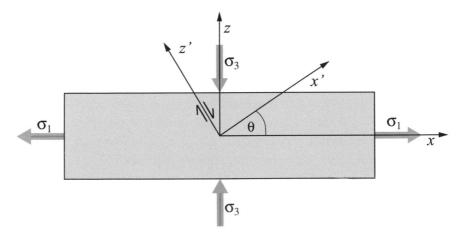

Figure 7.9 Layer of rock subjected to vertical lithostatic stress and extensional tectonic stress. Sliding will occur on an optimally oriented fault plane at angle θ with respect to the vertical.

In the previous chapter on elasticity, we discussed stress and strain in three dimensions as well as the development of a principal stress coordinate system where shear stresses are zero. Here we perform a stress tensor transformation in two dimensions by counterclockwise rotation through an angle θ. In the 2-D $x - z$

principal coordinate system, the stress tensor is simply

$$\begin{bmatrix} \sigma_1 & 0 \\ 0 & \sigma_3 \end{bmatrix}. \tag{7.18}$$

Transformation to the $x' - z'$ coordinate system involves a rotation

$$\mathbf{R} = \begin{bmatrix} \cos\theta & \sin\theta \\ -\sin\theta & \cos\theta \end{bmatrix}. \tag{7.19}$$

The tensor rotation is

$$\sigma' = \mathbf{R}\sigma\mathbf{R}^T \tag{7.20}$$

where

$$\sigma' = \begin{bmatrix} \sigma'_{xx} & \sigma'_{xz} \\ \sigma'_{xz} & \sigma'_{zz} \end{bmatrix}. \tag{7.21}$$

Exercise 7.2 Show that the components of the stress tensor in the rotated coordinate system are

$$\begin{aligned} \sigma'_{xx} &= \sigma_1\cos^2\theta + \sigma_3\sin^2\theta \\ \sigma'_{zz} &= \sigma_1\sin^2\theta + \sigma_3\cos^2\theta \\ \sigma'_{xz} &= (\sigma_3 - \sigma_1)\sin\theta\cos\theta \end{aligned} \tag{7.22}$$

Next consider the shear stress and normal stress on a plane perpendicular to the $x'-$ axis. This will be the optimally oriented fault plane where the frictional sliding will occur. In this case $\tau = \sigma'_{xy}$ so

$$\tau = \frac{(\sigma_3 - \sigma_1)}{2}\sin 2\theta \tag{7.23}$$

where we have used the trigonometric identity $2\sin\theta\cos\theta = \sin 2\theta$.

Exercise 7.3 Show that the normal stress is given by

$$\sigma_n = \frac{-(\sigma_3 + \sigma_1)}{2} + \frac{(\sigma_3 - \sigma_1)}{2}\cos 2\theta \tag{7.24}$$

You will need the trigonometric identity $\cos 2\theta = \cos^2\theta - \sin^2\theta$. The negative sign comes from the definition of the normal stress which is positive in compression whereas the stress convention used in this book is positive in extension.

Next note that Byerlee's law has the following form

$$\tau = S_o + f\sigma_n \tag{7.25}$$

where S_o is the cohesion of the rock and f is the coefficient of friction. We seek the angle θ that has the minimum shear stress to cause sliding on the fault. Equating the Byerlee shear stress with the shear stress in the rotated coordinate system we have

$$(\sigma_3 - \sigma_1) \sin 2\theta = 2S_o - f (\sigma_3 + \sigma_1) + f (\sigma_3 - \sigma_1) \cos 2\theta \qquad (7.26)$$

or

$$(\sigma_3 - \sigma_1) (\sin 2\theta - f \cos 2\theta) = 2S_o - f (\sigma_3 + \sigma_3) . \qquad (7.27)$$

Exercise 7.4 Use the computer algebra capabilities of MATLAB or another computer algebra program to show that the optimal angle is given by

$$\tan 2\theta = \frac{1}{f} \qquad (7.28)$$

and

$$\sigma_1 = \sigma_3 + 2 (S_o + f\sigma_3) \left[\left(1 + f^2\right)^{1/2} + f \right]. \qquad (7.29)$$

A plot of the shear stress τ versus normal stress σ_n as a function of the dip angle of the normal fault θ is shown as a Mohr circle in Figure 7.10. Also shown is Byerlee's law. The point where these functions intersect is the minimum stress needed to activate slip on the optimally oriented fault.

We can express Byerlee's law in terms of the principal stresses to arrive at

$$\begin{array}{ll} \sigma_1 = 4.67\sigma_3 & \sigma_3 < 113 \text{ MPa} \quad \theta = 24.8 \\ \sigma_1 = 3.12\sigma_3 + 176 \text{ MPa} & \sigma_3 > 113 \text{ MPa} \quad \theta = 29.5 \end{array} . \qquad (7.30)$$

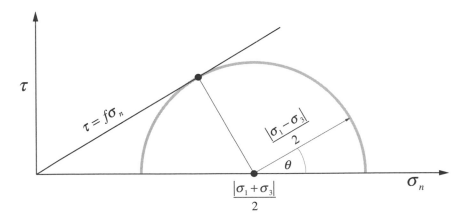

Figure 7.10 Mohr diagram showing the intersection point of Byerlee's law ($f = 0.6$) with a Mohr circle.

Now there are two cases to consider for the orientation of the principal stresses – horizontal extension and horizontal compression. For the extensional case the smaller principal stress is vertical so $\sigma_3 = \sigma_v = \rho g z$. The differential stress, horizontal minus vertical is

$$(\sigma_h - \sigma_v) = 0.786 \rho g z \qquad \rho g z < 530 \text{ MPa} \quad \theta = 24.8$$
$$(\sigma_h - \sigma_v) = 0.679 \rho g z + 56.7 \text{ MPa} \quad \rho g z > 530 \text{ MPa} \quad \theta = 29.5 \qquad (7.31)$$

Finally, we consider the compressional case where the largest principal stress is vertical so $\sigma_1 = \sigma_v = \rho g z$. The differential stress, vertical minus horizontal is

$$(\sigma_v - \sigma_h) = 3.67 \rho g z \qquad \rho g z < 113 \text{ MPa} \quad \theta = 65.2$$
$$(\sigma_v - \sigma_h) = 2.12 \rho g z + 176 \text{ MPa} \quad \rho g z > 113 \text{ MPa} \quad \theta = 60.5 \qquad (7.32)$$

Yield strength corresponding to frictional sliding on optimally oriented faults for extension and compression are plotted in Figure 7.11. The optimally-oriented faults are about three times stronger in compression than in extension. Including water at hydrostatic pressure in the faults reduced the normal stress and thus the yield strength.

Note that this theory provides the dip of an optimally oriented extensional (normal) fault of between 60.5 and 65.5° from a horizontal plane. When discussing outer rise normal faults in Chapter 9, the best fitting coefficient of friction is ~0.3 corresponding to a fault dip of 53°. An idealized frictionless normal fault would have a dip angle of 45°.

7.6.3 Ductile Deformation

As temperature increases with depth, rocks begin to undergo ductile flow in response to differential stress. Below the brittle-ductile transition, the stress needed to cause ductile flow becomes less than the stress needed to cause frictional sliding. The stress needed to produce ductile flow is sensitive to temperature, differential stress, mineral composition, grain size, water content, and strain rate and is relatively insensitive to pressure. Geological strain rates vary from 10^{-17} s^{-1} to 10^{-14} s^{-1}. Temperatures vary from 0 °C to the melting temperature of olivine ~1400 °C. Differential stresses of interest in geodynamics are typically greater than 10 MPa and less than 200 MPa. Under these conditions the dominant ductile deformation mechanism is dislocation creep. The flow law commonly used to describe dislocation flow in crystalline materials is a thermally activated power-law relation between strain rate $\dot{\varepsilon}$ and differential stress $(\sigma_1 - \sigma_3)$. The most commonly

Table 7.3. *Material parameters for dislocation creep.*

Rock	n	A	Q (J mol^{-1})
granite (wet)	1.9	7.9×10^{-16}	1.41×10^5
orthopyroxene (dry)	2.4	1.2×10^{-16}	2.93×10^5
olivine	3.0	7.0×10^{-14}	5.23×10^5

used relationship has the following form

$$\dot{\varepsilon} = A(\sigma_1 - \sigma_3)^n \exp\left(\frac{RT}{Q}\right) \tag{7.33}$$

where the parameters are

$$
\begin{array}{clc}
\dot{\varepsilon} & \text{strain rate} & 10^{-15}\ \text{s}^{-1} \\
A & \text{material constant} & \text{Pa}^{-n}\ \text{s}^{-1} \\
(\sigma_1 - \sigma_3) & \text{differential stress} & \text{Pa} \\
n & \text{stress exponent} & 2-3 \\
Q & \text{activation energy} & \text{J mol}^{-1} \\
R & \text{gas constant} & 8.314\ \text{J mol}^{-1}\ \text{K}^{-1} \\
T & \text{absolute temperature} & \text{K}
\end{array}
\tag{7.34}
$$

For this flow law, strain rate increases exponentially with increasing temperature and as a power of the differential stress. To further develop the yield strength envelope, this equation can be inverted for differential stress as

$$(\sigma_1 - \sigma_3) = \left(\frac{\dot{\varepsilon}}{A}\right)^{1/n} \exp\left(\frac{Q}{nRT}\right). \tag{7.35}$$

There are three laboratory-derived material constants that depend on rock type and grain size. For our simplified analysis we will just consider rock types related to the mantle (olivine) and end member continental crustal rocks ranging from wet granite to dry orthopyroxene (Table 7.3)

The overall yield strength versus depth including ductile flow is shown in Figure 7.11 for a geotherm corresponding to 100 Ma of cooling. Ocean lithosphere has crustal thickness of 6–7 km so the frictional sliding yield mechanism dominates through the crust and into the upper mantle. The solid curves show yield stress for frictional sliding of faults containing water at hydrostatic pressure while the dashed curves correspond to dry faults. The lithosphere is much stronger in compression than extension because when the horizontal compression exceeds the lithostatic

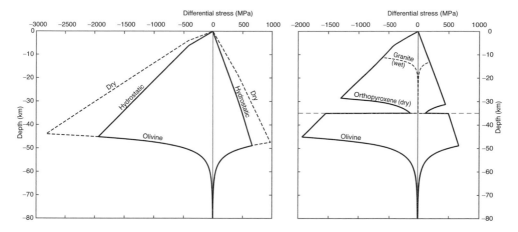

Figure 7.11 Yield strength versus depth for oceanic (left) and continental (right) lithosphere having a geotherm corresponding to 100 Ma old oceanic lithosphere according to the plate cooling model discussed in Chapter 5.

pressure the normal stress is increased. Similarly, extension causes unclamping of the faults so they can slide with less differential stress.

The yield strength versus depth for continental lithosphere is more complex. For a typical crustal thickness of 35 km, the lower crust deforms by ductile flow and it is highly dependent on the rock type, water content, and grain size. For this relatively cold, 100 Ma geotherm, the crust consisting of dry orthopyroxene has a weak basal crustal layer between depths of 30 and 35 km. In contrast, if the crust consists of wet granite, it will be weak in the depth range 10–35 km. Regardless of the rock type, the vertically integrated strength of continental lithosphere is always less than oceanic lithosphere of the same geotherm. This has several important implications for geodynamics:

1. Given the same horizontal driving force applied to a plate containing continental and oceanic crust, the continental lithosphere will deform first. This is evident in earthquake patterns where seismicity primarily occurs at the boundaries of the ocean lithosphere but is more diffuse in the continental plates.
2. Continental lithosphere having thick crust of say 50 km will be much weaker than continental lithosphere having normal 35 km crust. As proposed by Vink et al. (1984), continental rifts will prefer to propagate through thicker crust associated with mountains. This could help explain the Wilson Cycle where continental rifting preferentially occurs at suture zones.

3. Finally, as we will see in flexural analyses, the effective elastic thickness of
   continental lithosphere is less than the effective elastic thickness for the ocean
   lithosphere for a similar geotherm.

7.6.4 Strength vs. Age

To further illustrate the effects of temperature and crustal thickness on the strength
of the lithosphere, we have calculated YSE models for lithosphere having cooling
ages between 4 and 256 Ma (Figure 7.12). The ocean lithosphere has yield strength
that increases with age. The overall thickness-integrated strength in extension
increases almost linearly with age out to about 70 Ma where it begins to flatten
owing to the finite-thickness of the plate cooling model. The strength versus age

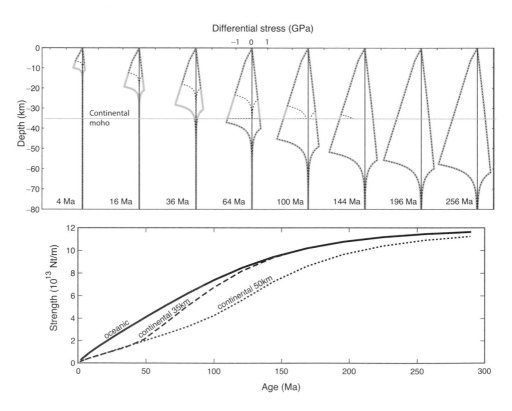

Figure 7.12 Oceanic (solid) and continental (dashed) yield strength envelope
models as a function of cooling age based on the plate cooling model. (upper)
Yield strength versus depth in extension (positive) and compression (negative).
(lower) Integrated yield strength for oceanic (solid) and continental lithosphere
having normal crustal thickness (dashed) and 50 km thick crust (short dash). Note
that for the same geotherm, continental lithosphere is always weaker than oceanic
lithosphere.

for continental lithosphere (dry orthopyroxene) is more complicated owing to the ductile yielding of the lower crust. As discussed previously, continental lithosphere is always weaker than oceanic lithosphere for the same thermal structure. Moreover, the strength depends on crustal thickness. For a 50 Ma geotherm, continental lithosphere is about half the strength of ocean lithosphere. At 100 Ma the continental lithosphere with 50 km crust is still half the strength of oceanic lithosphere. For very cold lithosphere, the continents and oceans have about equal strength. Finally, it is interesting to compare the integrated strength of the oceanic lithosphere with the plate driving forces. At 100 Ma, the integrated strength is 75×10^{12} N m^{-1} while the ridge push force is much smaller at 3.2×10^{12} N m^{-1}. In the last chapter we estimate the slab pull force at about $13 - 37 \times 10^{12}$ N m^{-1} at 80 Ma. Therefore the strength of the oceanic lithosphere usually exceeds the driving forces so the plates remain largely undeformed.

7.7 Exercises

Exercise 7.5 What is the average crustal stress needed to maintain the elevation of Tibet (5 km) with respect to the elevation of India (0 km — sea level)? Use the crust and mantle densities of 2800 kg m^{-3} and 3200 kg m^{-3}, respectively. Assume the crustal thickness under India is 30 km.

(a) Use Airy isostasy to solve for the crustal thickness of Tibet.
(b) Calculate the outward driving force using equation (7.7).
(c) Calculate the average stress as the ratio of driving force to total crustal thickness.

Exercise 7.6 An ice sheet of thickness D and density $\rho_I = 960$ kg m^{-3} is floating on an ocean of density $\rho_w = 1025$ kg m^{-3} under the force of gravity g. Assume the ice sheet is thin relative to its horizontal dimensions. See Figure 7.13.

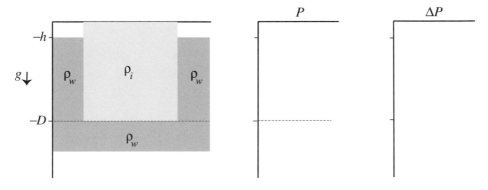

Figure 7.13

(a) Sketch the pressure versus depth for the ice and water, as well as the pressure difference versus depth.
(b) Derive an expression for the freeboard (i.e., height above sea level) of the ice sheet.
(c) Derive an expression for the outward driving force per unit length into the page caused by this density configuration

$$F_s = \int_{-D}^{0} \Delta P(z) \, dz$$

Check your results in a limiting case.

8

Flexure of the Lithosphere

This chapter is basically a supplement to *Geodynamics* (Turcotte and Schubert, 2014, Chapter 3). The results of the first derivation are the same as *Geodynamics*, equation (3.130), but rather than guessing the general solution, the solution is developed using Fourier transforms. The approach is similar to the solutions of the marine magnetic anomaly problem, the lithospheric heat conduction problem, the strike-slip fault flexure problem, and the flat-Earth gravity problem. In all these cases, we use the Cauchy integral theorem to perform the inverse Fourier transform. Later we'll combine this flexure solution with the gravity solution to develop the gravity-to-topography transfer function. Moreover, one can take this approach further, to develop a Green's function relating temperature, heat flow, topography, and gravity to a point heat source (e.g., Sandwell, 1982). In addition to the constant flexural rigidity solution found in the literature, we develop an iterative solution to flexure with spatially variable rigidity.

Before going over these notes, read *Geodynamics*, Section 3.9, on the development of moment versus curvature for a thin elastic plate.

The loading problem is illustrated in Figure 8.1. We start with a simple line source, but the solution method also applies to a point source. Of course, the point source Green's function can be convolved with an arbitrary load distribution to make the solution completely general; we'll do this later. The vertical force balance for flexure of a thin elastic plate floating on the mantle is described by the following differential equation.

$$\underbrace{\frac{d^2}{dx^2}\left(D(x)\frac{d^2w}{dx^2}\right)}_{\substack{\text{flexural}\\\text{resistance}}} + \underbrace{F\frac{d^2w}{dx^2}}_{\substack{\text{end}\\\text{load}}} + \underbrace{\Delta\rho g w}_{\substack{\text{restoring}\\\text{force}}} = \underbrace{q(x)}_{\substack{\text{vertical}\\\text{load}}} \tag{8.1}$$

Table 8.1.

Parameter	Definition	Value/Unit
$w(x)$	deflection of plate (positive down)	m
$D(x) = \dfrac{Eh(x)^3}{12(1-\nu^2)}$	flexural rigidity	N m
h	elastic plate thickness	m
F	end load	N m^{-1}
q	vertical load	N m^{-2}
$\Delta\rho$	density $(\rho_m - \rho_w)$ for ocean	2200 kg m^{-3}
g	acceleration of gravity	9.82 m s^{-2}
E	Young's modulus	6.5×10^{10} Pa
ν	Poisson's ratio	0.25

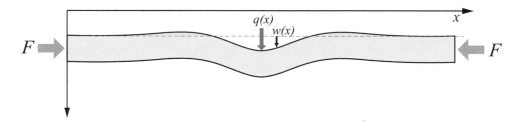

Figure 8.1 Thin elastic plate subjected to vertical load $q(x)$ and end load F. The plate responds with a vertical displacement $w(x)$.

The parameters are defined in Table 8.1.

8.1 Constant Flexural Rigidity, Line Load, No End Load

Under these assumptions, the differential equation and boundary conditions become

$$D\frac{\mathrm{d}^4 w}{\mathrm{d}x^4} + \Delta\rho g w = V_o\delta(x)$$

$$\lim_{|x|\to\infty} w(x) = 0 \quad \text{and} \quad \lim_{|x|\to\infty}\frac{\mathrm{d}w}{\mathrm{d}x} = 0. \tag{8.2}$$

Take the Fourier transform of the differential equation, where the forward and inverse transforms are now defined as

$$F(k) = \int_{-\infty}^{\infty} f(x)\,e^{-ikx}\,\mathrm{d}x \quad f(x) = \frac{1}{2\pi}\int_{-\infty}^{\infty} F(k)\,e^{ikx}\,\mathrm{d}k \tag{8.3}$$

where the wavenumber is now $2\pi/\lambda$ instead of the usual $1/\lambda$. The derivative property is now $\Im[dw/dx] = ik\Im[w]$. The Fourier transform of the differential equation is

$$Dk^4 W(k) + \Delta\rho g W(k) = V_o \tag{8.4}$$

and the solution for plate deflection is simply

$$W(k) = \left[k^4 + \frac{4}{\alpha^4}\right]^{-1} \frac{V_o}{D} \tag{8.5}$$

where the flexural parameter α is (see Turcotte and Schubert, 2014, equation (3.127))

$$\alpha^4 = \frac{4D}{\Delta\rho g}. \tag{8.6}$$

Now take the inverse Fourier transform of equation (8.5).

$$w(x) = \frac{V_o}{2\pi D} \int_{-\infty}^{\infty} \frac{e^{ikx}}{\left(k^4 + \frac{4}{\alpha^4}\right)} dk \tag{8.7}$$

As in the other solutions, we find the poles in the denominator of equation (8.7) and integrate around the poles.

$$\left(k^4 + \frac{4}{\alpha^4}\right) = \left(k^2 + \frac{2i}{\alpha^2}\right)\left(k^2 - \frac{2i}{\alpha^2}\right)$$

$$\left(k^4 + \frac{4}{\alpha^4}\right) = \left(k - \frac{1+i}{\alpha}\right)\left(k - \frac{-1+i}{\alpha}\right)\left(k - \frac{1-i}{\alpha}\right)\left(k - \frac{-1-i}{\alpha}\right) \tag{8.8}$$

See also Figure 8.2.

First consider the case for $x > 0$. To match the boundary conditions at infinity, we want $\text{Im}(k) > 0$. Thus, we close the integration in the upper half of the plane and apply the Cauchy residue theorem

$$\oint \frac{f(z)}{z - z_o} dz = i2\pi f(z_o). \tag{8.9}$$

The relevant poles are

$$k = \frac{1+i}{\alpha} \quad \text{and} \quad k = \frac{-1+i}{\alpha}. \tag{8.10}$$

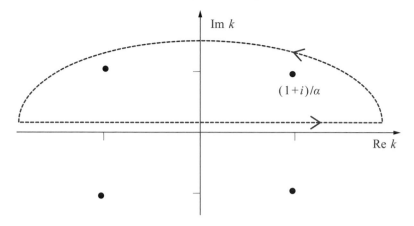

Figure 8.2 Location of poles in the complex plane. The integration path from $-\infty$ to ∞ can be closed in the upper or lower hemisphere.

The solution is

$$w(x) = \frac{V_o}{2\pi D} 2\pi i \left[\frac{\alpha^3 e^{i\left(\frac{1+i}{\alpha}\right)x}}{(1+i+1-i)(1+i-1+i)(1+i+1+i)} \right.$$

$$\left. + \frac{\alpha^3 e^{i\left(\frac{-1+i}{\alpha}\right)x}}{(-1+i-1-i)(-1+i-1+i)(-1+i+1+i)} \right]. \tag{8.11}$$

After some simplification, this becomes

$$w(x) = \frac{V_o \alpha^3}{8D} e^{-x/\alpha} \left[\frac{e^{ix/\alpha}}{(1+i)} + \frac{e^{-ix/\alpha}}{(1-i)} \right]. \tag{8.12}$$

This can be rewritten in terms of $\cos(x/\alpha)$ and $\sin(x/\alpha)$. Also, we know that the solution should be symmetric about $x = 0$. The final result for positive x matches (Turcotte and Schubert, 2014, equation (3.130)).

$$w(x) = \frac{V_o \alpha^3}{8D} e^{-x/\alpha} \left[\cos(x/\alpha) + \sin(x/\alpha) \right] \tag{8.13}$$

The important parameters and length scales in this solution are:

α	flexural parameter
$\sqrt{2}\pi\alpha$	flexural wavelength
$x_o = 3\pi\alpha/4$	distance to the first zero crossing.

For the case of a broken plate the solution is

$$w(x) = \frac{V_o \alpha^3}{4D} e^{-x/\alpha} \cos(x/\alpha). \qquad (8.14)$$

This is also the form used to model plate bending at a subduction zone. The general solution to this 2-D flexure problem has four terms:

$$w(x) = Ae^{-x/\alpha} \sin(x/\alpha) + Be^{-x/\alpha} \cos(x/\alpha)$$
$$+ Ce^{x/\alpha} \sin(x/\alpha) + De^{x/\alpha} \cos(x/\alpha). \qquad (8.15)$$

When solving for flexure in a plate with two or more step variations in flexural rigidity such as fracture zone flexure (Exercise 9.5), one needs all four terms.

Figure 8.3 shows the solution for the continuous plate where the maximum flexure is normalized to one. The solution for a broken plate is shown in Figure 8.4. Note that for the same downward force, the amplitude of the broken plate is twice the amplitude of the continuous plate. We will also see this in the next section where we use a variability rigidity formulation to simulate a broken plate.

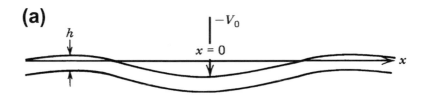

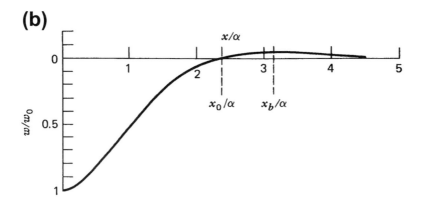

Figure 8.3 Deflection of a thin elastic plate under a line load. From *Geodynamics* (Turcotte and Schubert, 2014).

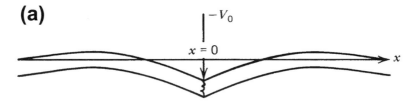

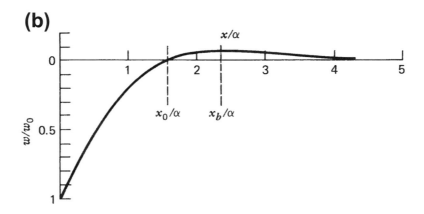

Figure 8.4 Deflection of a broken thin elastic plate under a line load. From *Geodynamics* (Turcotte and Schubert, 2014).

8.2 Variable Flexural Rigidity, Arbitrary Line Load, No End Load

For this case, we need to solve a linear differential equation, but with a variable coefficient. This will involve an iteration scheme in the Fourier transform domain where the first iteration is basically equation (8.5) above. See the original derivation in Sandwell (1984). The differential equation and boundary conditions are

$$\frac{d^2}{dx^2}\left(D(x)\frac{d^2 w(x)}{dx^2}\right) + \Delta\rho g w(x) = q(x)$$

(8.16)

$$\lim_{|x|\to\infty} w(x) = 0 \quad \text{and} \quad \lim_{|x|\to\infty} \frac{dw}{dx} = 0$$

where $D(x)$ is now the spatially variable flexural rigidity, $w(x)$ is the deflection of the plate, and $q(x)$ is the applied load. It is assumed that D and w are band-limited functions, so that their Fourier transforms exist. The functions D and w can be written as:

$$D(x) = \frac{1}{2\pi} \int_{-\infty}^{\infty} D(s) e^{isx} \, ds$$

$$(8.17)$$

$$w(x) = \frac{1}{2\pi} \int_{-\infty}^{\infty} W(r) e^{irx} \, dr.$$

Upon substitution of these expressions for D and w into the first term of equation (8.16) and differentiating under the integral, the following is obtained.

$$\frac{1}{(2\pi)^2} \int_{-\infty}^{\infty} \int_{-\infty}^{\infty} (r+s)^2 r^2 D(s) \, W(r) \, e^{i(s+r)x} \, dr \, ds + \Delta\rho g w(x) = q(x) \qquad (8.18)$$

The Fourier transform of equation (8.18) is

$$\frac{1}{(2\pi)^2} \int_{-\infty}^{\infty} \int_{-\infty}^{\infty} (r+s)^2 r^2 D(s) W(r) \int_{-\infty}^{\infty} e^{i(s+r-k)x} \, dx \, dr \, ds + \Delta\rho g W(k) = Q(k).$$

$$(8.19)$$

By making use of the definition of the delta function

$$\frac{1}{2\pi} \int_{-\infty}^{\infty} e^{i(s+r-k)x} \, dx = \delta\Big[r - (k-s)\Big] \qquad (8.20)$$

performing the integral with respect to r, and using the band-limited property of $D(s)$ (i.e., $D(s) = 0|s| > \beta$), equation (8.19) reduces to a Fredholm integral equation

$$\frac{k^2}{2\pi} \int_{-\beta}^{\beta} D(s) \, W(k-s)(k-s)^2 \, ds + \Delta\rho g W(k) = Q(k). \qquad (8.21)$$

Notice that if the flexural rigidity is constant, $D(x) = D_o$, then $D(s) = 2\pi D_o \delta(s)$. For this case, the solution for the plate deflection for an arbitrary load is

$$W(k) = \big[D_o k^4 + \Delta\rho g\big]^{-1} Q(k). \qquad (8.22)$$

Now consider the more general case of spatially variable flexural rigidity

$$D(s) = D'(s) + 2\pi D_o \delta(s). \qquad (8.23)$$

Inserting equation (8.23) into equation (8.21) and rearranging terms yields

$$W(k) = \left[D_o k^4 + \Delta\rho g\right]^{-1} \left[Q(k) - \frac{k^2}{2\pi} \int_{-\beta}^{\beta} D'(s)W(k-s)(k-s)^2 \, ds \right]. \quad (8.24)$$

The plate deflection $W(k)$ appears on both sides of equation (8.24), so there is no closed form solution for $W(k)$. However, if the variations in flexural rigidity D' are small compared with the mean value of flexural rigidity D_o, then this equation can be solved by successive approximation. The original derivation in Sandwell (1984) provides the necessary requirement for convergence, but a numerical illustration is also useful. This variable-rigidity flexure approach has also been extended to 3-D (Garcia et al., 2014).

Figure 8.5 is a numerical example of flexure of a plate with a sharp reduction in plate thickness at the origin. The upper curve compares the flexure of a continuous plate (solid curve, analytic solution of equation (8.13)) to the Fourier transform solution of equation (8.2) (dashed curve). The lower plot is a comparison of the

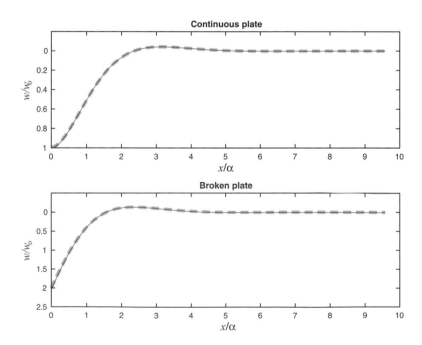

Figure 8.5 (upper) Flexure of a plate with constant flexural rigidity. The Fourier transform (dashed) and the analytic solution ($e^{-x/\alpha}(\cos x/\alpha + \sin x/\alpha)$) show good agreement. (lower) Flexure of a broken plate. The iterative Fourier transform solution (dashed) and analytic solution ($2e^{-x/\alpha}\cos x/\alpha$) show good agreement.

analytic solution to flexure of a broken plate (solid curve, (Turcotte and Schubert, 2014, equation (3.140))) to the numerical iterative solution of equation (8.24) (dashed curve). For this case, the thickness of the plate at the origin was reduced by 95%. This approximates the broken plate solution.

8.3 Stability of Thin Elastic Plate under End Load

Consider the original differential equation (8.1) and now assume constant flexural rigidity with a delta function line load and uniform compressive end load F. The line load by itself will cause flexure of the plate, as given in equation (8.13). For sufficiently large end load, this initial deflection will become amplified and the plate will buckle. Here we develop formulas for the critical end load when the amplification becomes unbounded. In addition, the buckling will occur at a particular wavelength λ_c. Under these assumptions, the differential equation becomes

$$D\frac{d^4 w}{dx^4} + F\frac{d^2 w}{dx^2} + \Delta\rho g w(x) = V_o \delta(x).$$ (8.25)

The Fourier transform of the differential equation is

$$Dk^4 W(k) - Fk^2 W(k) + \Delta\rho g W(k) = V_o$$ (8.26)

and the solution for plate deflection is

$$W(k) = V_o \left[Dk^4 - Fk^2 + \Delta\rho g\right]^{-1}.$$ (8.27)

The deflection of the plate is singular when the denominator goes to zero. We note that the denominator is a polynomial, so we can use the quadratic formula to search for the zeros of this equation. We can write this as a quadratic equation in k^2.

$$k^2 = \frac{F \pm \sqrt{F^2 - 4D\Delta\rho g}}{2D}$$ (8.28)

We know the wavenumber must be a real number and this only occurs when

$$F \geqslant \sqrt{4D\Delta\rho g}.$$ (8.29)

When F is equal to this critical value, we call this the critical end load F_c. The wavelength when this occurs is the critical wavelength given by

$$\lambda_c = 2\pi \left[\frac{D}{\Delta\rho g}\right]^{1/4}$$ (8.30)

where D is the flexural rigidity given by

$$D = \frac{Eh^3}{12(1 - v^2)}$$

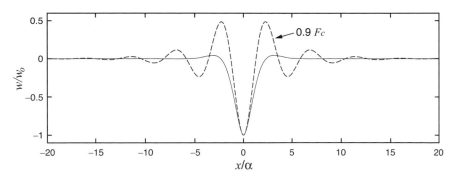

Figure 8.6 (solid) Normalized flexure of a continuous plate under a line load. (dashed) Normalized flexure with line load and end load. The plate begins to buckle at the flexural wavelength.

and h is the elastic plate thickness. Note this wavelength λ_c is also equal to the flexural wavelength discussed in Section 8.1.

An example of the flexure due to a line load on a thin elastic plate is shown in Figure 8.6. When no end load is applied, the flexure follows the analytic solution for a continuous plate given in equation (8.13). When an end load is applied the plate begins to buckle. This example has an end load $F = 0.9F_c$.

Next, consider an example of buckling of oceanic lithosphere that is 50 km thick. The density contrast $\Delta\rho = (\rho_m - \rho_w)$ and other parameters are given in Table 8.1. The parameters of interest are the average stress at the ends of the plate $\sigma = F_c/h$ and the buckling wavelength λ_c. For this case, the values are 4.9 GPa and 475 km. From the analysis of the yield strength envelope, it is clear that this level of stress cannot be sustained by even the strongest oceanic lithosphere. Therefore, the elastic buckling model is not relevant for the Earth. One must consider the inelastic properties of the lithosphere to understand the response to large end loads.

8.4 Exercises

Exercise 8.1 Continental yield strength envelope model. The continental yield strength has been described as a jelly sandwich consisting of a weak layer (jelly) between two strong layers (bread). The flexural rigidity of a single strong layer is

$$D = \frac{Eh^3}{12(1 - v^2)}. \tag{8.31}$$

(a) What is the flexural rigidity of two strong layers, each of thickness $h/2$, that are not bonded along their common interface?

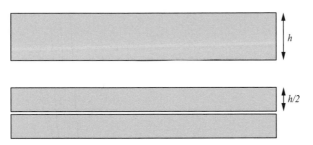

Figure 8.7 (solid) (top) Single plate of thickness h. (bottom) Two plates each of thickness $h/2$ are not bonded at their interface, so they act independently under a bending moment.

(b) What is the effective elastic thickness for this layered case (bottom diagram in Figure 8.7)?

Exercise 8.2 Write a MATLAB program to generate the two flexure curves in Figure 8.6. Start with the code used in Exercise 2.9.

Exercise 8.3 (a) Consider an ice shelf of thickness h (150 m) and density ρ_i (980 kg m^{-3}) floating on an ocean of density ρ_w (1025 kg m^{-3}). The ice starts at a cold uniform temperature of T_o ($-22\,^\circ$C) as it flows out onto the ocean and then is warmed at its base to a temperature of seawater T_w ($-2\,^\circ$C). The ice has enough time to equilibrate so the geotherm is approximately linear with depth and is

$$T(z) = (T_w - T_o)\frac{z}{h} + T_o. \tag{8.32}$$

Show that the thermal bending moment is

$$M_T = \frac{\gamma E (T_w - T_o)}{12} h^2 \tag{8.33}$$

where E (1.6×10^9 Pa) is Young's modulus and γ ($5.5 \times 10^{-5}\,^\circ$C^{-1}) is the linear coefficient of thermal expansion. Assume that the depth-integrated end load is zero.

(b) Derive the following formula for the deflection of the plate as a function of distance from the edge of the ice shelf. Hint: see equation (3.151) in *Geodynamics*.

$$w(x) = \frac{\alpha^2 M_T}{2D} e^{-(x/\alpha)} \left[\cos(x/\alpha) - \sin(x/\alpha)\right]. \tag{8.34}$$

(c) Plot the deflection of the plate as a function of distance from the edge of the ice shelf. How does this compare with the topography of the rampart and moat shown in Figure 2 of Scambos et al. (2005)?

9

Flexure Examples

This chapter provides practical examples of flexural models applied to structures in the lithospheres of Earth and Venus. The models are all basically solutions to the thin-plate flexure equation, with a variety of surface loads, sub-surface loads, and boundary conditions. Both gravity and topography data are used to constrain the models. We'll see in chapter 17 that gravity data provide important constraints on the topography of the Moho. Figures and captions from various sources are provided on the following pages. The excellent book by Watts (2001) provides a much more thorough and extensive survey of oceanic and continental flexures.

The features discussed below include:

Seamounts undersea volcanoes loading the oceanic lithosphere
Trenches plate bending at subduction zones on Earth and Venus
Fracture Zones flexure that accumulates due to the differential subsidence across
        an oceanic fracture zone.

In addition, there are nine exercises at the end of the chapter to explore a wider variety of published flexure models including ice shelves and rift flank uplifts. Students can select a topic of interest, reproduce the published results, and provide a critical analysis of each paper.

9.1 Seamounts

Seamounts are undersea volcanoes found mainly in the deep ocean. They were originally defined by Menard (1964) as "isolated elevation from the seafloor with a circular or elliptical plan, at least 1 km of relief, comparatively steep slopes and a relatively small summit area". They are produced in four tectonic settings:

near mid-ocean spreading ridges; in plate interiors over upwelling mantle plumes (hot spots); in areas of intraplate extension; and in island-arc convergent settings (Wessel, 2001). Although they represent a tiny fraction of the volume of the oceanic crust, they are scientifically important for a number of reasons: the geochemistry of seamounts provides a window into the upper mantle; the loading of seamounts on the lithosphere provides estimates of lithospheric thickness and rheology; ocean currents/tides are impeded by seamounts causing ocean mixing and upwelling; and seamounts provide habitats for a variety of sea life (*Mountains in the Sea* (Staudigel et al., 2010)). Since only about 20% of the seafloor has been mapped with multibeam echo sounders, only a fraction of seamounts have been identified. Here we focus on understanding the thickness and rheology of the lithosphere using relatively large seamounts as impulsive sources of flexure. The lithospheric cooling models, discussed in previous chapters, predict that seamount-induced flexure will reflect the elastic thickness of the lithosphere at the time the seamount formed (e.g. Watts, 1978).

In Chapter 8 we calculated the flexure due to a line load on a thin elastic plate. Of course, this line load is not well suited to a seamount, which is more like a point load. We will see in Chapters 16 and 17 that flexure and gravity models are easily calculated for loads of arbitrary shape using Fourier transform methods. The approach is to take a 2-D Fourier transform of the load (i.e., bathymetry), use the thin elastic plate flexure equation (17.5) to compute the topography of the Moho; then use the solution to Poisson's equation (16.33) to calculate the gravity anomaly produced by both the topography and Moho interfaces. The model gravity can be compared with the observed gravity to find parameters that provide the smallest misfit. The main unknown parameters are the elastic thickness and the crustal density.

As an example, we model flexure caused by large seamounts of the Foundation seamount chain in the South Pacific (Figure 9.1). This area of seafloor has complete multibeam bathymetry coverage so the bathymetry and gravity data sets are independent. The gravity modeling is performed with Generic Mapping Tools (GMT) (Wessel et al., 2019). The topography and gravity data are from matching global grids having 1 minute spatial resolution (Tozer et al., 2019; Sandwell et al., 2019). The computer scripts to make the computations follow in Table 9.1. The analysis starts with a topography grid that has boundaries that extend >300 km beyond all sides of the grid shown in Figure 9.1. This extension is needed to avoid Fourier edge effects, and also include wavelengths much longer than the expected flexural wavelength. The isostatic parameters of a smaller area or individual seamount can be estimated by selecting a sub-area to evaluate the rms misfit between the observed and model gravity.

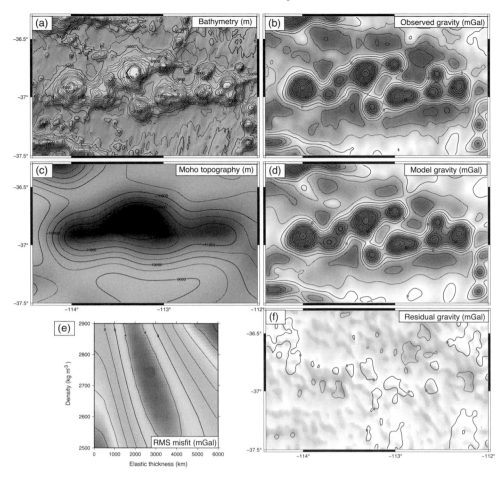

Figure 9.1 Gravity/bathymetry flexure model of the Foundation seamounts.
(a) Bathymetry of the eastern section of the Foundation seamounts closest
to the Pacific Antarctic ridge on seafloor younger than 3 Ma. (b) Free-air
gravity anomaly derived from satellite altimetry. (c) Moho topography based on
bathymetric loading of a thin elastic plate ($h = 2500$ m, $\rho_c = 2750$ kg m^{-3},
$\rho_m = 3300$ kg m^{-3}) (d) Model gravity based on gravity contributions from the
bathymetry and Moho. (e) Rms misfit between observed and model gravity with
mean removed versus elastic thickness and crustal density. (f) Difference between
observed and model gravity. Modeling was done using Generic Mapping Tools
(GMT) (Wessel et al., 2019). (For a color version of this figure, please see the
color plate section.)

9.1.1 Elastic Thickness Versus Age

Flexural analyses have been performed by numerous investigators at hundreds of
locations around the oceans. Compilations such as shown in Figure 9.2 are used
to better understand the thickness and strength of the cooling oceanic lithosphere.

Table 9.1. *GMT computer script and data used to compute gravity/topography flexure models anywhere on the Earth.*

```
foundations.csh
#  compare observed and model gravity in the area of the Foundation seamounts
#  gather the global grids of topography and gravity anomaly
wget -q ftp://topex.ucsd.edu/pub/global_topo_{1}min/topo_{2}0.1.nc
wget -q ftp://topex.ucsd.edu/pub/global_grav_{1}min/grav_{2}9.1.nc
#  specify and cut the full and rms regions
set region = "-R-120/-106/-42/-32"
set rms_region = "-R-114.4/-112/-37.5/-36.3"
gmt grdcut topo_{2}0.1.nc -Gtopo.grd $region
gmt grdcut grav_{2}9.1.nc -Ggrav.grd $region
#  compute the optimal model
set rhoc = "2750"
set Te   = "2500"
isostatic.csh topo.grd 1025 $rhoc 3300 $Te 6500 g_iso.grd
#  cut to the rms region
gmt grdcut g_iso.grd $rms_region -Gg_iso_cut.grd
gmt grdcut grav.grd $rms_region -Ggrav_cut.grd
#  subtract the isostatic anomaly from the gravity and compute stdev
gmt grdmath grav_cut.grd g_iso_cut.grd SUB = g_diff.grd
gmt grdinfo -L2 g_diff.grd  | grep stdev
# clean up
rm topo.grd grav.grd g_diff.grd g_iso.grd

isostatic.csh
# script to compute a flexural isostatic correction for topography
if ($#argv < 7) then
 echo " "
 echo "Usage: isostatic.csh topo.grd rhow rhoc rhom Te Dc g_iso.grd"
 echo " "
 echo "Example: isostatic.csh topo.grd 1025 2700 3300 5000 7000 g_iso.grd"
 echo " "
 exit 1
endif
alias MATH 'set \!:1 = `echo "\!:3-$" | bc -l`'
#   set the densities and elastic thickness
set RW = $2
set RC = $3
set RM = $4
set TE = $5
MATH RCRW = $RC - $RW
# compute the mean ocean depth and moho depth
set DT = `gmt grdinfo -L2 $1 | grep mean | awk '{if($3 < 0) print(-$3); else print (100)}'`
MATH DM = $6 + $DT
#  compute the gravity from the topography using 4 terms in the Parker expansion
gmt gravfft $1 -D$RCRW -E4 -Gg_topo.grd
#  compute the gravity due to the compensation and the moho topography
gmt gravfft $1 -T$TE/$RC/$RM/$RW+m -E1 -Z$DM  -Gg_comp.grd
#  add the two contributions
gmt grdmath g_topo.grd g_comp.grd ADD = sum.grd
#  filter the model to match the spectrum of the data
gmt grdfilter sum.grd -Fg14 -D2 -G$7
#  cleanup
rm g_topo.grd g_comp.grd sum.grd
```

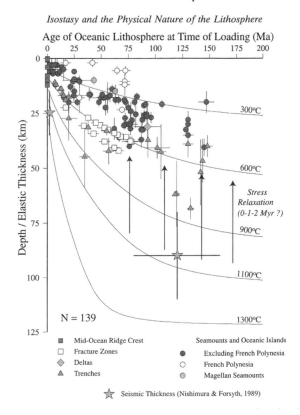

Figure 9.2 Elastic thickness versus the age of the lithosphere at the time of loading compiled by Watts (2001)

Flexural analyses at spreading ridges generally find low elastic thickness associated with hot and young lithosphere. Many seamount flexure studies show an increase in the elastic plate thickness with the age of the lithosphere at the time of loading. The base of the elastic later has a temperature between 300 and 600 °C. However, some seamounts in French Polynesia have anomalously low elastic thickness that is not well understood in terms of simple plate cooling. Plate bending at ocean trenches provides very clear flexural signals but their elastic thickness versus age is highly scattered. One factor not accounted for in these flexural studies is the possible non-linear relationship between bending moment and curvature. The simplest linear models have a single slope (i.e., the flexural rigidity D) to the moment vs. curvature relationship as described in Chapter 8. However, this simple relationship breaks down when the plates are bent beyond their elastic limit. We'll explore these non-linear effects in the next section on trench flexure where the plates are bent at angles up to 10 degrees prior to subduction.

9.2 Trenches

Deep ocean trenches are the locations where the ocean lithosphere is bent downward beginning its subduction into the mantle. This bending produces very large topography and gravity anomalies (Figure 9.3) so these are excellent locations to study lithospheric flexure (Figure 9.4). In addition, since the flexure is occurring now, we know the age of the lithosphere when it was flexed.

In this section we show examples of combined topography and gravity flexure analyses at numerous trenches (Figure 9.5). As discussed above, the elastic thickness derived from trench data does not show a clear correlation with lithospheric age. We also highlight the fractures and normal faults that occur on the outer trench

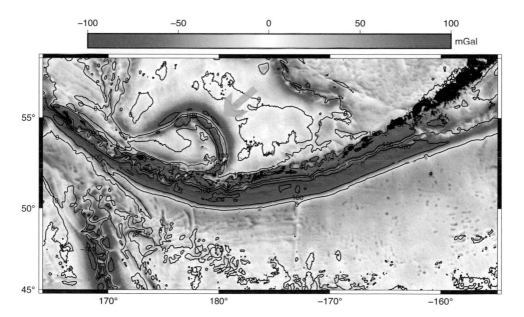

Figure 9.3 Gravity anomaly of the Aleutian Trench (100 mGal contour interval). The deepest part of the trench has large negative anomalies (-200 mGal). The outer rise south of the trench has smaller positive anomalies (20–40 mGal). Note the extinct trench marked by a light gray arrow has a large gravity anomaly demonstrating that trench flexures reflect bending of an elastic plate rather than a viscous plate since the viscous flexure would decay in a few million years. (For a color version of this figure, please see the color plate section.)

TOPOGRAPHY MODEL

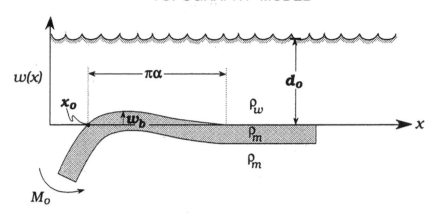

GRAVITY MODEL

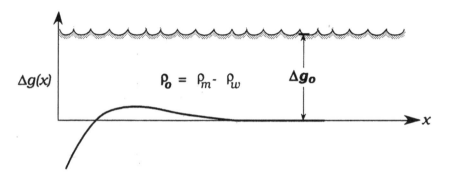

Figure 9.4 Schematic representation of topography and gravity at an ocean trench. The model parameters include the elastic thickness h, the first zero crossing outboard of the trench x_0, the width of the outer rise $\pi\alpha$, the mean depth far from the trench d_0, the regional gravity far from the trench g_0, and the density of the lithosphere ρ_m (Levitt and Sandwell, 1995).

wall. Both observations illustrate that the lithosphere is bent beyond its elastic limit at trenches. This is not unexpected since the bending is permanent and continues deeper into the Earth reaching dip angles of 45 degrees or more from the horizontal.

9.2.1 Moment Saturation at Trenches and Outer Rise Normal Faults

The elastic thickness versus age derived from seamount and trench flexure shows considerable scatter (Figure 9.2) suggesting that the elastic thickness is highly

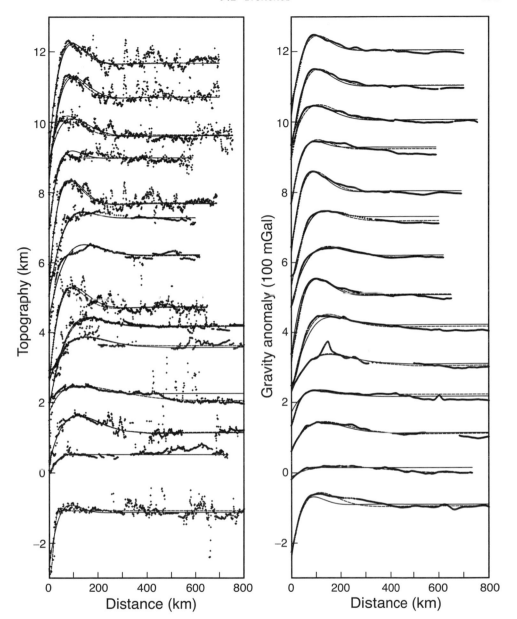

Figure 9.5 Topography and gravity data for 14 of 117 profiles modeled in the Levitt and Sandwell (1995) study. Profiles are shifted vertically for presentation. Data are shown as tiny crosses. Solid lines correspond to the best fit model using both topography and gravity and dashed lines correspond to the best fit model using just topography (left) or gravity (right).

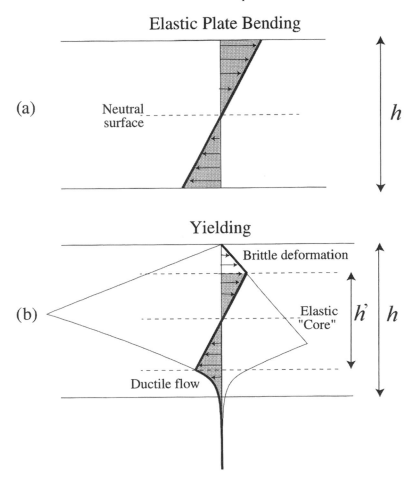

Figure 9.6 (a) Stress versus depth in a bent thin elastic plate has extension near the surface and compression at depth. (b) Stresses cannot exceed the yield strength of the plate resulting in brittle fracture near the surface and ductile flow at depth. The elastic core is thinner than the original elastic thickness because of yielding. Modified from Watts (2001).

variable with age. Another interpretation, shown in Figure 9.6, is that the elastic thickness measured by flexure h' represents the elastic core of the plate which is always less than the thickness h predicted from the plate cooling model (McNutt and Menard, 1978). This happens because the upper part of the plate undergoes brittle fracture and the lower part of the plate deforms plastically. Rather than use this effective elastic thickness as a measure of plate strength, it is better to use the maximum flexural bending moment since it accounts for these non-linear effects from yielding. We call this the saturation bending moment M_s and it is the integral of the yield stress $\sigma_Y(z)$ shown as the thick solid line in Figure 9.6(b) times the distance from the neutral surface.

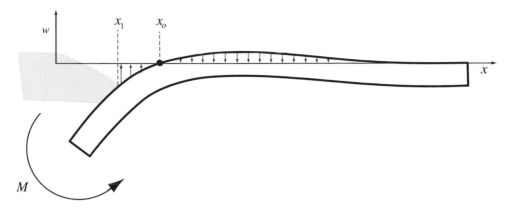

Figure 9.7 Flexural topography caused by a bending moment applied to the subducted lithosphere.

$$M_s = \int_{-h/2}^{h/2} \sigma_Y(z) z \, dz \qquad (9.1)$$

The bending moment at a trench cannot exceed this saturation bending moment. (Note that the depth to the neutral surface should be adjusted to ensure that the overall end load on the plate is zero.)

One can directly measure the bending moment at a trench axis and adjust the rheology model of the lithosphere to make sure it does not exceed the saturation moment. If we assume that the flexural topography at a trench is entirely supported by a bending moment, then the moment can be measured as shown in Figure 9.7. The moment is simply the topographic restoring stress $g \Delta \rho w(x)$ times the moment arm $(x - x_o)$ integrated from the trench axis to infinity

$$M = g \Delta \rho \int_{x_1}^{\infty} w(x)(x - x_o) \, dx \leq M_s \qquad (9.2)$$

In theory one could measure this moment directly from the topography but it is difficult to select the zero level. A small error in the zero level results in a very large error in the moment because there will be non-zero topography when the moment arm is large.

A more robust approach to measuring the maximum moment at the trench axis is to first fit a flexure model to bathymetry and then use that model to calculate the maximum bending moment at the trench axis (Levitt and Sandwell, 1995). Gravity anomaly data are also used in the fitting (Figures 9.4 and 9.5) to help

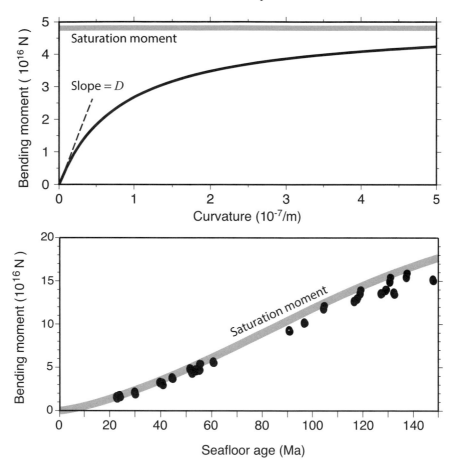

Figure 9.8 (upper) Bending moment versus curvature for a yield strength envelope model having a 0.3 coefficient of friction with hydrostatic pore pressure and a
ductile flow law. The temperature versus depth is based on the Parsons and Sclater
(1977) plate cooling model at and age of 60 Ma. At low curvature ($<10^{-8}$) there is
a linear relation between moment and curvature having a slope equal to the flexural
rigidity corresponding to the mechanical thickness of the plate. As the curvature
increases, the moment versus curvature flattens and eventually saturates. (lower)
Maximum bending moment at the trench axis versus age for 24 trenches in the
Pacific basin (modified from Garcia et al., 2019). The thick grey curve shows the
maximum bending that can be sustained as a function of age. Note that all of these
subducting plates are nearly moment saturated.

constrain both the shape of the flexure and also the zero level of the topography
far from the trench. Note that gravity is less noisy than bathymetry because of
the attenuation of the short wavelengths by upward continuation from the ocean
floor to the ocean surface. Recent results are shown in Figure 9.8 for 24 trench

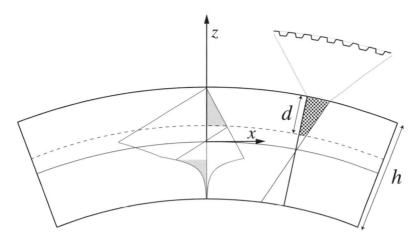

Figure 9.9 Schematic diagram showing a plate of thickness h that is bent beyond its elastic limit. According to the YSE formulation, fractures will extend to a depth d. We assume that the upper zone of inelastic deformation is incompressible (stippled). The depth-averaged extension in that zone is accommodated by faulting and graben formation on the surface (modified from Garcia et al., 2019).

flexure areas around the Pacific ocean basin (Garcia et al., 2019). The observed maximum bending moment increases by more than a factor of 10 between ages of 20 and 140 Ma. Also shown is the saturation bending moment which is calculated using a YSE formulation from Goetze and Evans (1979) but using parameter values from Mueller and Phillips (1995) combined with the temperatures from the plate cooling model (Parsons and Sclater, 1977, Pacific model). The upper brittle layer is relatively weak with a 0.3 coefficient of friction and hydrostatic pore pressure. The observed bending moment is close to the saturation bending moment so the elastic core is close to zero thickness. This explains why the effective elastic thickness from trench flexure studies is generally smaller than the expected mechanical thickness of the plate (e.g., depth to 600–800 °C isotherm).

As shown in Figures 9.8, the plate is nearly moment saturated at the trench axis, so the upper part of the plate, where strength is controlled by brittle fracturing, should have faults that extend from the surface to nearly half the plate thickness. With a few assumptions we can use the model to predict the vertical offset on the outer rise faults. The model for the outer rise fractures is shown in Figures 9.9. The plate of initial thickness h has a curvature of $\partial^2 w/\partial x^2$ where x is along the direction of maximum curvature. The amount of strain ε is related to the curvature and the distance from the neutral surface by $\varepsilon = -z\partial^2 w/\partial x^2$ (Turcotte and Schubert, 2014). The strain is purely elastic near the neutral surface, but there is brittle deformation from the surface down to a depth d. We assume that this yielding region is incompressible, so that horizontal extension in the upper brittle layer is accommodated by vertical deformation. The deformation will appear on the surface as down-dropped blocks. The average vertical offset on the surface is then equal to the horizontal strain integrated through the thickness of the upper deformed layer. The result is

$$\Delta h = \frac{1}{2} \frac{\partial^2 w}{\partial x^2} d \, (h - d) \, . \tag{9.3}$$

For a plate having an initial 50 km thickness, a faulting depth of 20 km, and a typical magnitude of curvature of $4 \times 10^{-7} \text{m}^{-1}$, the vertical relief will be 120 m. If the ratio of undisturbed surface (horst) to down-dropped blocks (graben) is 1/2, then the vertical relief will be 240 m.

We can compare this model for the formation of horst and graben topography on the outer trench walls where there is high resolution multibeam sonar data. Examples are shown in Figures 9.10 and 9.11 for the Tonga and Japan Trenches, respectively. The horst and graben topography grows toward the trench axis to an amplitude of about 400 m, in basic agreement with the model prediction. However, this simple model predicts larger topography than is observed and the onset of the observed horst and graben topography is closer to the trench than the model predicts. These misfits could be due to a number of processes such as thermoelastic pre-stress (Garcia et al., 2019), and elastic processes that would reduce the amplitude of the model which assumes incompressible material.

9.2.2 Venus Trench

In the final example of this trench section we explore these same flexure, fracture, and moment saturation concepts for features on the planet Venus. We observe several features on Venus having arcuate planform with horizontal dimensions similar to trenches on the Earth (Figures 9.12 and 9.13) (e.g., (Schubert and Sandwell, 1995)). Moreover, these trench features have arcuate fractures similar to the horst and graben structures on the seaward walls of Earth trenches (Figure 9.14). While this suggests there has been lithospheric subduction on Venus, we do not have the confirming evidence such as Benioff zones showing subduction of lithosphere to at least 700 km into the Earth's mantle. Moreover, the total length of trench-like features on Venus is only 30% of the length of the trenches on Earth. Also, if

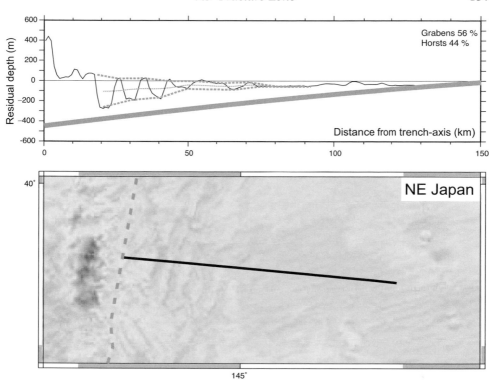

Figure 9.10 Outer rise fractures in NE Japan: (upper) Residual bathymetry across the seaward wall of the trench and outer rise. The dashed grey lines traverse the top of horst blocks and the base of grabens. The distance between these surfaces is used to estimate the depth of grabens. The dotted grey line is equidistant between the top (horst) and bottom (graben) surfaces and is used to estimate the proportion of horsts and grabens. Estimated fault throw for the YSE model described above (thick grey curve) has more amplitude than the observations. (lower) Residual swath bathymetry. The dashed grey and solid black lines mark the trench axis and the location of the profile, respectively (modified from Garcia et al., 2019).

subduction is prevalent today on Venus there should be significant evidence for lithospheric spreading or widespread lithospheric extension. Nevertheless, the topography and gravity of these features is highly suggestive of subduction on Venus and this hypothesis could possibly be tested with an additional spacecraft survey of topography and surface change.

9.3 Fracture Zone

A fracture zone is a scar in the lithosphere produced by a transform fault as shown in Figure 9.15. The transform fault has active strike-slip motion between

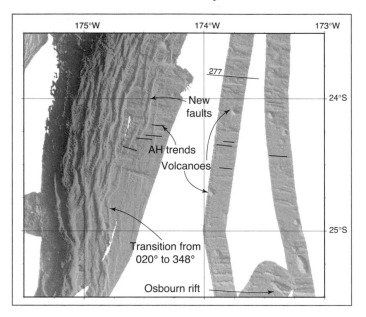

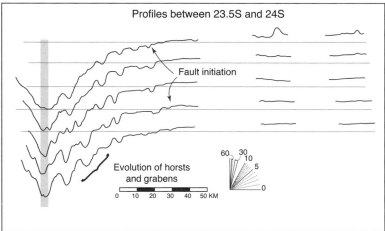

Figure 9.11 High resolution multibeam bathymetry data reveal the development of horst and graben structures on the outer trench wall of the Tonga trench (Massell, 2002). Seaward of these fractures the abyssal hill fabric runs in an east–west direction which is perpendicular to the bending fractures.

the two ridges but the two plates begin to move at the same rate outboard of the ridge-transform intersection (RTI). At the RTI, the lithosphere on the young side A is hot and weak while the lithosphere on the old side A' has cooled and strengthened. When the two sides begin to move at the same rate they fuse, forming

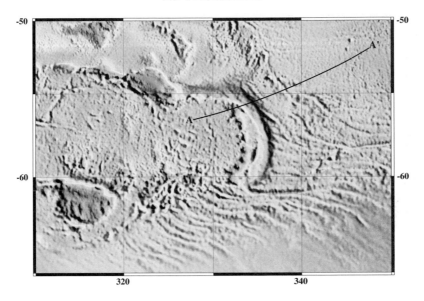

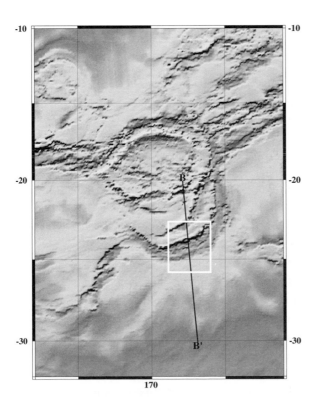

Figure 9.12 (upper) Shaded bathymetry map of the Sandwich trench, South Atlantic. (lower) Shaded topography of Latona Corona, Venus. The maps are plotted at the same horizontal and vertical scale (Sandwell and Schubert, 1992). Profiles A–A′ and B–B′ are profiles shown in Figure 9.13. The white box outlines the high resolution topography shown in Figure 9.14. (For a color version of this figure, please see the color plate section.)

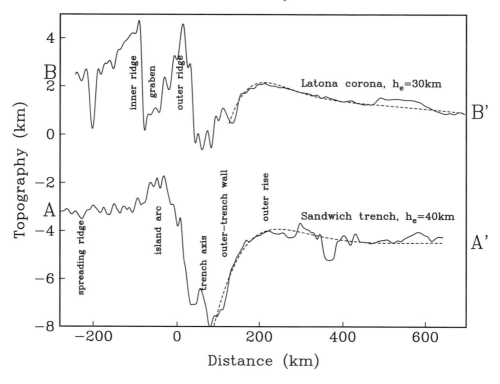

Figure 9.13 Profiles across the Sandwich Trench (A–A') and Latona Corona (B–B') show similar flexural wavelength and amplitude (Sandwell and Schubert, 1992). Dashed curves are fits of flexure models.

a single strong fracture zone. The young side is cooling and subsiding faster than the older side and if they remain bonded, the initial depth difference at the RTI may be frozen in place. To maintain the initial depth offset at the RTI, and also to accommodate the shrinking depth difference across the FZ, the plates must flex. The thermal evolution problem was given as an exercise in Chapter 5. In this section we show a schematic of the evolution of a fracture zone (Figure 9.15) and provide comparisons of the thermal subsidence/flexure model predictions with bathymetry profiles across the Mendocino and Pioneer FZ's (Figure 9.16). The solution to the flexure model is given as an exercise at the end of this chapter. Several studies have pointed out that the plates may have partial slip when the differential subsidence rate is high just outboard of the RTI, so this is an area of active research that will require detailed multibeam sonar surveys of more fracture zones.

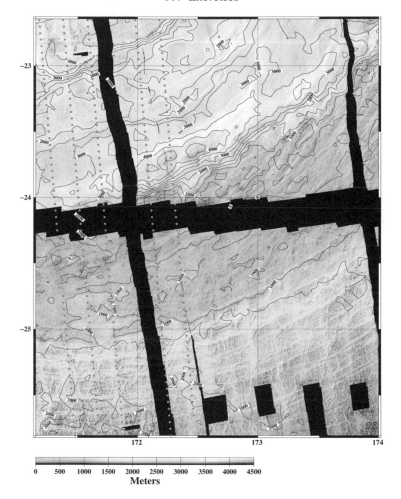

Figure 9.14 Topography (200 m contour interval) of a segment of southern Latona Corona superimposed on a Magellan SAR image reveals the relationship between circumferential fractures and the major ridges, trenches, and scarps (Schubert and Sandwell, 1995). These fractures on the outer trench wall are similar to the fractures seen outboard of the Tonga trench in Figure 9.11, suggesting that they formed when the lithosphere outboard of Latona Corona was flexed beyond its elastic limit.

9.4 Exercises

Exercise 9.1 Show that Equation 9.2 is true by integrating the solution for the trench flexure case, equation (3.142) in *Geodynamics*.

$$w(x) = w_o e^{-x/\alpha} \cos \frac{x}{\alpha} \tag{9.4}$$

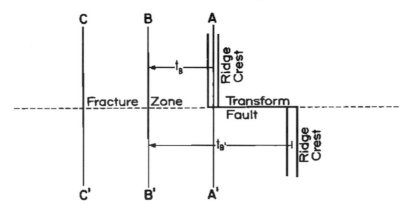

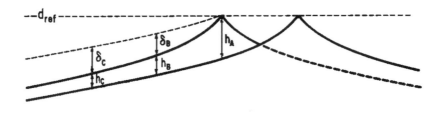

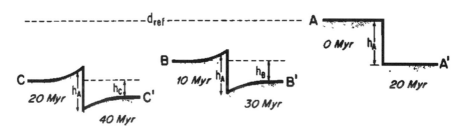

Figure 9.15 Evolution of a fracture zone (Sandwell and Schubert, 1982a). (Top)
Spreading ridges offset by a transform fault. The age of offset across the FZ is
$t_{B'} - t_B$. (Center) The h's are the differences in ocean floor depth between locations
far to either side of the FZ. The initial height of the scarp at the FZ is h_A. If the FZ
does not slip, the scarp height must remain constant. The constancy of scarp height
and the decrease in h with age cause the lithosphere in the vicinity of the FZ to
bend. The flexural amplitude δ_B is the difference between h_A and h_B. Similarly,
$\delta_c = h_A - h_C$. (Bottom) Sketch of bathymetry along profiles A–A', B–B', and
C–C' illustrating the lithospheric flexure described above.

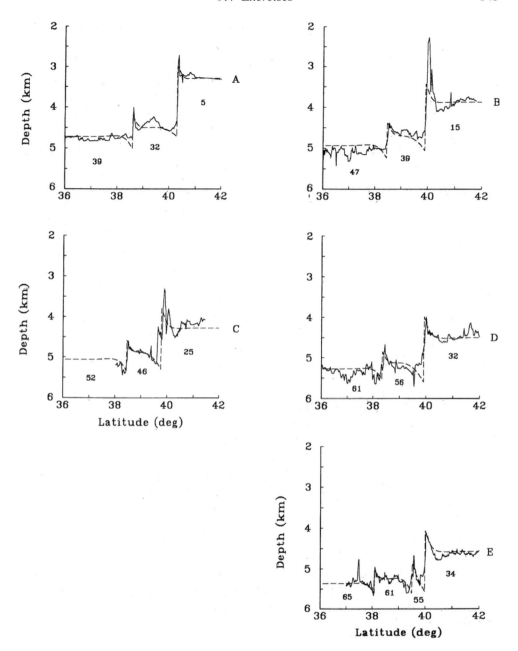

Figure 9.16 Comparisons between theoretical bathymetric profiles computed from flexure models assuming no slip on the FZs (dashed lines) and the observed bathymetric profiles A–E (solid lines) (Sandwell and Schubert, 1982a). The asymmetric flexure predicted by the model across each FZ is a consequence of the increase in flexural wavelength with age. The apparent tilt in the bathymetry between the Pioneer and Mendocino FZs occurs because the flexural wavelength is greater than their separation distance.

Compare this with the moment computed from the curvature

$$M(x) = D\frac{\partial^2 w}{\partial x^2}. \tag{9.5}$$

You will need the relationship between α and D given in Equation 8.6.

Exercise 9.2 Ice Shelf Flexure Discuss the tidal flexure model and derive equation (3) in the paper by Vaughan (1995). Generate the curve shown in Figure 5. Explain the overall findings of the paper.

Exercise 9.3 Seamount Flexure Discuss the seamount loading flexure model and derive equation (11) of Banks et al. (1977). How would one calculate a model for a load of arbitrary shape? Use GMT or MATLAB to generate the flexure for a Gaussian-shaped seamount.

Exercise 9.4 Trench Flexure Discuss the trench flexure problem and derive the solution given in equation (2) of Caldwell et al. (1976). Reproduce the graphs shown in Figure 3. Explain the overall findings of the paper.

Exercise 9.5 Fracture Zone Flexure Discuss the fracture zone flexure problem and derive the solution given in equation (11) of Sandwell and Schubert (1982a). Calculate the topography and stress across a single fracture zone with different flexural rigidities on either side (simple case, no time dependence, no lateral heat conduction). Explain the overall findings of the paper.

Exercise 9.6 Flexure on Venus Discuss the Venus Flexure problem and why it is important. Derive equation (2) in Johnson and Sandwell (1994). Derive equation (10) from equations (7) and (8). When might it be more appropriate to use a ring load rather than a bar load? Explain the overall findings of the paper in terms of the geothermal gradient on Venus.

Exercise 9.7 Outer Rise Yield Strength Discuss why it is important to consider the finite yield strength of the lithosphere when modeling flexure at subduction zones. Discuss equations (3) and (12) in McNutt and Menard (1982). Discuss the difference between the effective elastic thickness and the mechanical thickness.

Exercise 9.8 Rift Flank Uplift Why do the flanks of rifts go up? Reproduce Figure 3 in Brown and Phillips (1999). Discuss the equation (11) and Figure 5 in Weissel and Karner (1989).

Exercise 9.9 Lake Loading Discuss the lake loading flexure problem and its effect on the San Andreas Fault. Derive equations (2) and (3) in Luttrell et al. (2007). Reproduce the thin-plate plots in Figure 4. Explain the overall findings of the paper.

10

Elastic Solutions for Strike-Slip Faulting

(References: Weertman and Weertman, 1966; Savage and Burford, 1973; Cohen, 1999)

This chapter provides the mathematical development for the deformation and strain pattern due to an infinitely long, strike-slip fault in an elastic half space. The notes are similar to Sections 8.6–8.9 in *Geodynamics* (Turcotte and Schubert, 2014). While we follow the overall theme of Chapter 8, we'll deviate in two respects. First, we'll use a coordinate system with the z-axis pointed upward, to be consistent with the other chapters on gravity, magnetics, and heat flow. Second, we'll develop the solution from first principles using the Fourier transform approach. This approach does not explicitly use dislocations (e.g. Segall, 2010) but simulates dislocations using body force couples following Steketee (1958) and Burridge and Knopoff (1964).

10.1 Interseismic Strain Buildup

The first objective is to derive an expression for the surface displacement $v(x)$ and surface strain $\delta v/\delta x$ for the model shown in Figure 10.1. A constant velocity V is applied at an elastic half space. There is a fault in the half space with locked and creeping sections.

The approach will be as follows:

1. Develop the force balance from basic principles.
2. Establish the line-source Green's function for an elastic full space.
3. Establish the screw-dislocation Green's function for an elastic full space.
4. Use the method of images to construct a half-space solution.

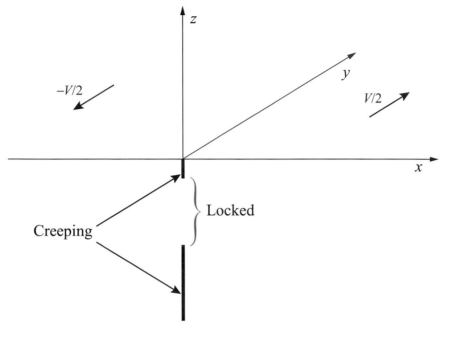

Figure 10.1

5. Integrate the line sources to develop the solutions found in the literature.
6. Compute the geodetic moment accumulation rate for an arbitrary slip distribution.
7. Use the inclined fault plane model.
8. Look at MATLAB examples.

10.1.1 Force Balance

Consider the forces acting on the infinitely long (y-direction) square rod depicted in Figure 10.2. The body force per unit volume of rod must be balanced by tractions on the sides of the rod.

The equation for this force balance is

$$\left[\tau_{xy}\left(x + \delta x\right) - \tau_{xy}\left(x\right)\right] \delta y\, \delta z + \left[\tau_{zy}\left(z + \delta z\right) - \tau_{zy}\left(z\right)\right] \delta x\, \delta y = b\left(x, z\right)\, \delta z\, \delta y\, \delta z \tag{10.1}$$

where τ_{xy} and τ_{zy} are the shear tractions on the side and top of the box, respectively, and $b(x, y)$ is the body force that depends only on x and z. Dividing through by $\delta x\, \delta y\, \delta z$ and taking the limit as all three go to zero, we arrive at:

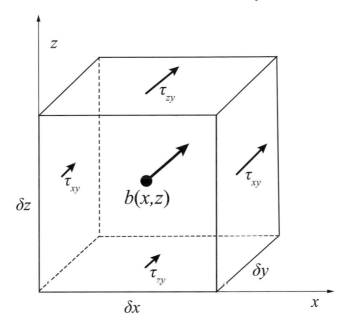

Figure 10.2

$$\frac{\partial \tau_{xy}}{\partial x} + \frac{\partial \tau_{zy}}{\partial z} = b\,(x, z)\,. \tag{10.2}$$

Given the following relationship between stress and displacement, the differential equation reduces to Poisson's equation

$$\tau_{xy} = \mu \frac{\partial v}{\partial x}$$
$$\tau_{zy} = \mu \frac{\partial v}{\partial z} \tag{10.3}$$

$$\frac{\partial^2 v}{\partial x^2} + \frac{\partial^2 v}{\partial z^2} = \frac{1}{\mu} b\,(x, z) \tag{10.4}$$

where μ is the shear modulus and v is the displacement in the y-direction.

10.1.2 Line-Source Green's Function

We can generate the solution to an arbitrary source distribution by first developing the line-source Green's function. Consider a line source at a depth of $-a$. The differential equation is

$$\frac{\partial^2 v}{\partial x^2} + \frac{\partial^2 v}{\partial z^2} = \frac{A}{\mu}\delta(x)\delta(z + a) \tag{10.5}$$

where A is the source strength having units of force/length, or force/length/time if this will represent an interseismic velocity. The boundary conditions for this second-order, partial differential equation are that v must vanish as both $|x|$ and $|z|$ go to infinity. The two-dimensional forward and inverse Fourier transforms are defined as

$$F(\mathbf{k}) = \int_{-\infty}^{\infty} \int_{-\infty}^{\infty} f(\mathbf{x})\, e^{-i2\pi(\mathbf{k}\cdot\mathbf{x})} d^2\mathbf{x}$$

$$f(\mathbf{x}) = \int_{-\infty}^{\infty} \int_{-\infty}^{\infty} F(\mathbf{k})\, e^{i2\pi(\mathbf{k}\cdot\mathbf{x})} d^2\mathbf{k}$$

(10.6)

where $\mathbf{k} = (k_x, k_y)$ and $\mathbf{x} = (x, y)$. Take the two-dimensional Fourier transform of the differential equation (10.5)

$$-(2\pi)^2 \left(k_x^2 + k_z^2\right) V(\mathbf{k}) = \frac{A}{\mu} e^{i2\pi k_z a}$$

(10.7)

so the solution in the Fourier domain is

$$V(\mathbf{k}) = \frac{-A e^{i2\pi k_z a}}{\mu(2\pi)^2 \left(k_x^2 + k_z^2\right)}.$$

(10.8)

Now we need to take the inverse Fourier transform with respect to k_z and make sure the solution goes to zero as $|z|$ goes to infinity. The integral is

$$V(k_x, z) = \frac{-A}{\mu(2\pi)^2} \int_{-\infty}^{\infty} \frac{e^{i2\pi k_z(z+a)}}{\left(k_x^2 + k_z^2\right)} dk_z.$$

(10.9)

First consider the case $k_x > 0, z + a > 0$. We can factor the denominator and recognize that the integrand will vanish for large positive z if we close the contour in the upper hemisphere.

$$V(k_x, z) = \frac{-A}{\mu(2\pi)^2} \oint \frac{e^{i2\pi k_z(z+a)}}{\left(k_z + ik_x\right)\left(k_z - ik_x\right)} dk_z$$

(10.10)

See Figure 10.3.

From the Cauchy integral formula, we know that for any analytic function the following holds for a counterclockwise path surrounding the pole.

$$\oint \frac{f(z)}{z - z_o} dz = i2\pi f(z_o)$$

(10.11)

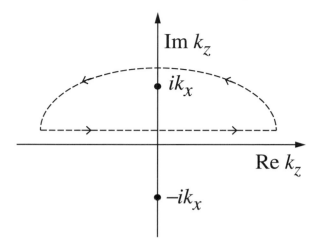

Figure 10.3

In this case with the pole at ik_x, the result is simply

$$V(k_x, z) = \frac{-i2\pi A}{\mu 4\pi^2} \frac{e^{-2\pi k_x(z+a)}}{i2k_x} = \frac{-A}{2\mu} \frac{e^{-2\pi k_x(z+a)}}{2\pi k_x}. \tag{10.12}$$

Next consider $k_x < 0, z + a > 0$. In this case, we must close the integration path in the lower hemisphere to satisfy the boundary conditions; during the integration the only contribution will be from the $-ik_x$ pole. The overall result is to replace k_x by $|k_x|$.

$$V(k_x, z) = \frac{-A}{2\mu} \frac{e^{-2\pi |k_x|(z+a)}}{2\pi |k_x|} \tag{10.13}$$

Note this is exactly the same functional form as the heat flow solution in Chapter 2. The Green's function is the inverse cosine transform of equation (10.13), or $\ln(r^2)$. The final result is

$$v(x, z) = \frac{-A}{4\pi \mu} \ln\left[x^2 + (z+a)^2\right]. \tag{10.14}$$

10.1.3 Screw Dislocation for Line Source Green's Function

In order to produce a fault plane with strike-slip displacement, we need to construct a line-source screw dislocation. This can be accomplished by abutting equal but opposite line source dislocations, as shown in Figure 10.4.

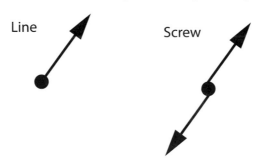

Figure 10.4

A simple way of constructing the screw source is to take the derivative of the line source Green's function in a direction normal to the fault plane. So, we need to develop the Green's function for the following differential equation.

$$\nabla^2 v_{screw} = \delta(z+a)\Big[\delta(x+dx) - \delta(x)\Big] \Big/ dx = \delta(z+a)\frac{\partial}{\partial x}\delta(x) \qquad (10.15)$$

To do this, we take the derivative of the line source Green's function in equation (10.14).

$$v_{screw}(x,z) = \frac{-A}{4\pi\mu}\frac{\partial}{\partial x}\ln\Big[x^2 + (z+a)^2\Big] = \frac{-Ax}{2\pi\mu}\Big[x^2 + (z+a)^2\Big]^{-1} \qquad (10.16)$$

So the Green's function for a line-source screw dislocation at depth is

$$v_{screw}(x,z) = \frac{-A}{2\pi\mu}\frac{x}{\Big[x^2 + (z+a)^2\Big]}. \qquad (10.17)$$

10.1.4 Surface Boundary Condition: Method of Images

The surface boundary condition is that the shear stress τ_{zy} must be equal to zero, but the full-space result provides a non-zero result. This boundary condition will be satisfied if we place an image source at $z = a$. When the combined source and image are evaluated at the surface $z = 0$, the result is to double the strength of the Green's function.

$$v(x,z) = \frac{-A}{2\pi\mu}x\left\{\Big[x^2 + (z+a)^2\Big]^{-1} + \Big[x^2 + (z-a)^2\Big]^{-1}\right\}$$

$$(10.18)$$

$$v(x,0) = \frac{-A}{\pi\mu}\frac{x}{\Big[x^2 + a^2\Big]}$$

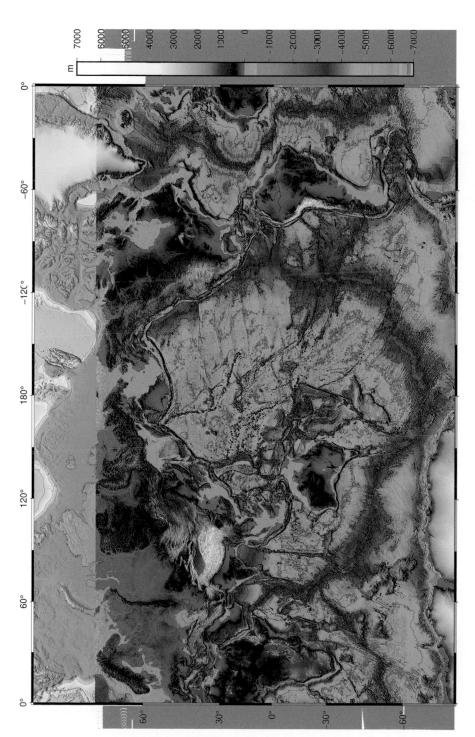

Figure 1.1 Topography of the Earth based on a global compilation of land data (SRTM and other sources) and ocean data from a combination of sparse ship soundings and marine gravity anomalies derived from satellite altimetry (Smith and Sandwell, 1997; Tozer et al., 2019).

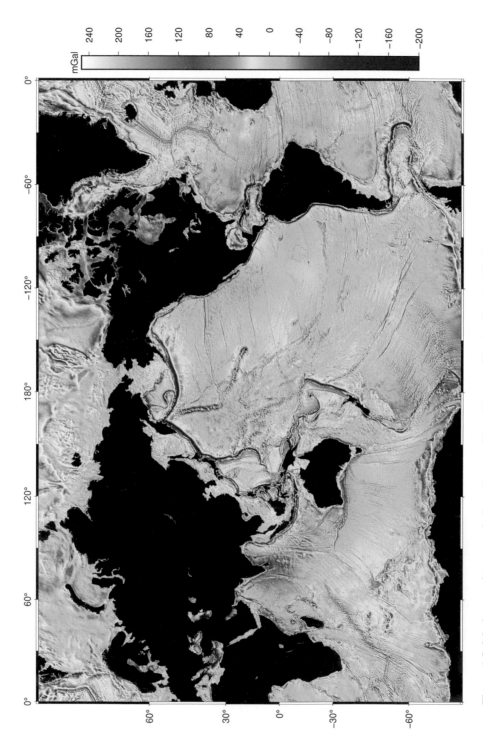

Figure 1.3 Marine gravity anomaly based on satellite altimetry (Sandwell et al., 2014).

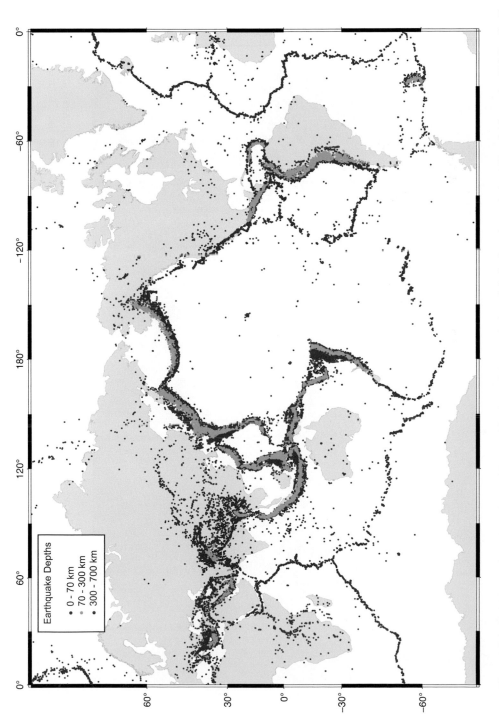

Figure 1.4 Well-located earthquakes with magnitude >5.1 reveal the global plate boundaries (Engdahl et al., 1998). Shallow earthquakes (0–70 km: red) provide clear definition of the boundaries of the oceanic plates, but have a more diffuse distribution on the continents. Intermediate (70–300 km: green) and deep (300–700 km: blue) earthquakes occur in subduction zones and are the primary evidence for lithospheric subduction to at least 670 km.

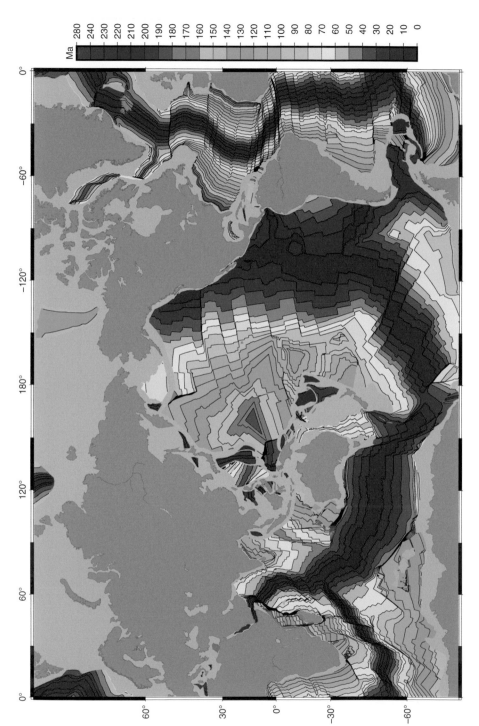

Figure 1.5 Seafloor age (Müller et al., 1997; Seton et al., 2020) based on identified magnetic anomalies and relative plate reconstructions along trends identified in satellite altimeter measurements of marine gravity. Ages in the Cretaceous quiet zone (64–127 Ma), the Jurassic (145–200 Ma) and older have poor control.

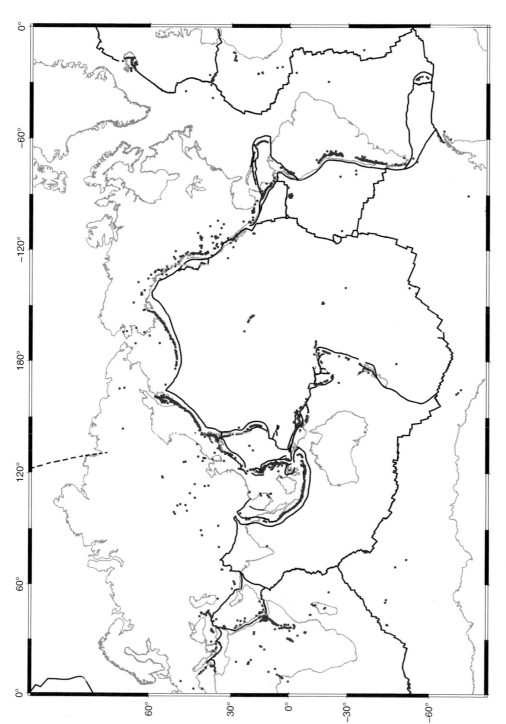

Figure 1.6 Plates, plate boundaries, and subaerial Quaternary volcanoes. The dozen or so tectonic plates are separated by spreading centers, transform faults, and subduction/thrust faults. The majority of active or recently active volcanoes (Siebert and Simkin, 2002) are associated with convergent plate boundaries.

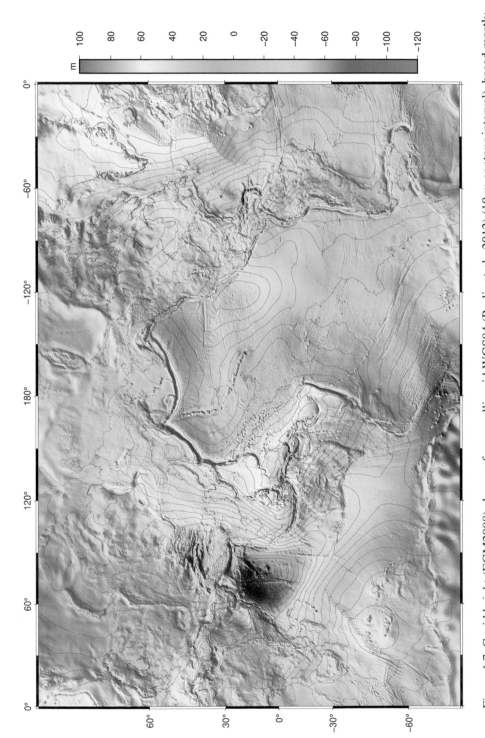

Figure 1.7 Geoid height (EGM2008) above reference ellipsoid WGS84 (Pavlis et al., 2012) (10 m contour interval), based mostly on satellite tracking data and some terrestrial gravity anomaly measurements. Unlike topography, seismicity, and age shown in the other maps, the geoid is poorly correlated with surface tectonics, except in areas where mature lithosphere has subducted in the western Pacific.

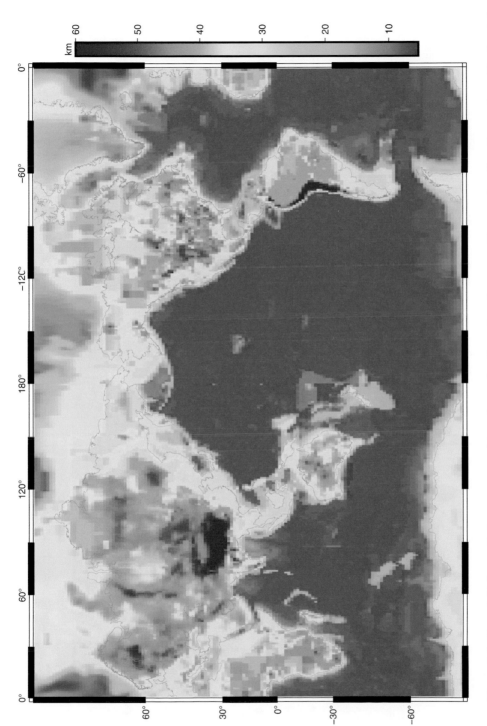

Figure 1.8 Crustal thickness (1 degree averages) based on refraction seismology as well as receiver function analyses (Laske et al., 2013). Gravity anomalies were used to estimate crustal thickness in areas with no constraints.

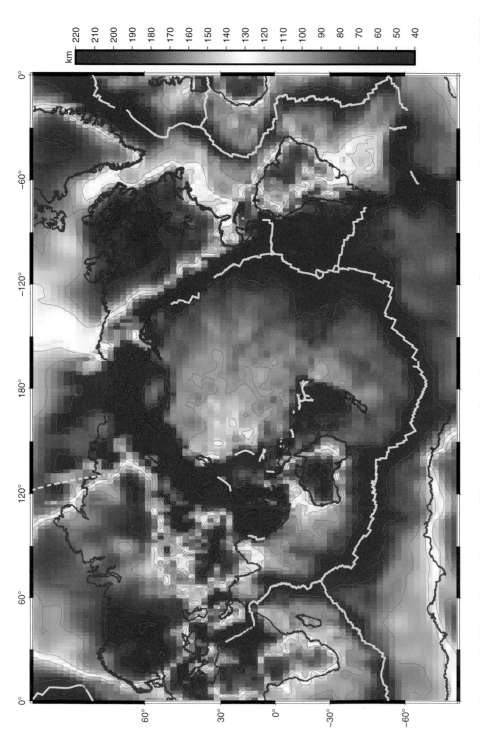

Figure 1.9 Lithospheric thickness (20 km contour interval) based on surface wave tomography (Priestley et al., 2018). Present-day spreading ridges are shown as yellow line.

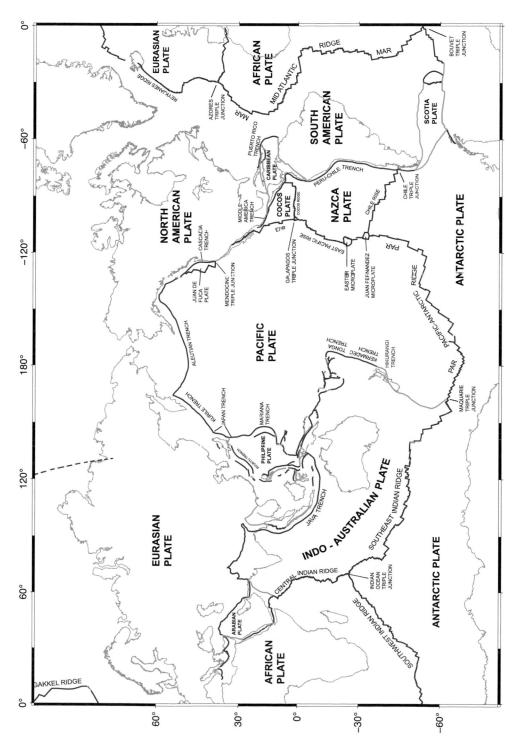

Figure 3.5 Names of major plates and plate boundaries modified from Fowler (1990). Ridges are red, transforms are green, and trenches are blue.

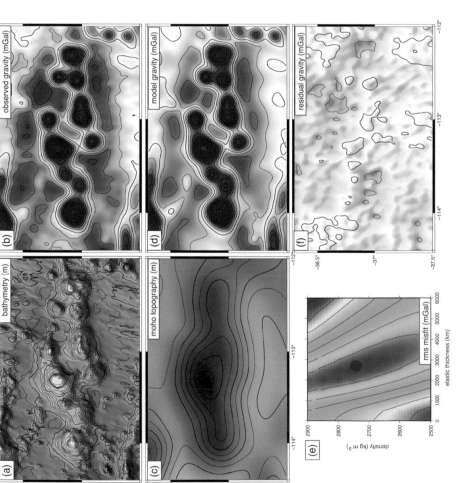

Figure 9.1 Gravity/bathymetry flexure model of the Foundation seamounts. (a) Bathymetry of the eastern section of the Foundation seamounts closest to the Pacific Antarctic ridge on seafloor younger than 3 Ma. (b) Free-air gravity anomaly derived from satellite altimetry. (c) Moho topography based on bathymetric loading of a thin elastic plate ($h = 2500$ m, $\rho_c = 2750$ kg m^{-3}, $\rho_m = 3300$ kg m^{-3}) (d) Model gravity based on gravity contributions from the bathymetry and Moho. (e) Rms misfit between observed and model gravity with mean removed versus elastic thickness and crustal density. (f) Difference between observed and model gravity. Modeling was done using Generic Mapping Tools (GMT) (Wessel et al., 2019).

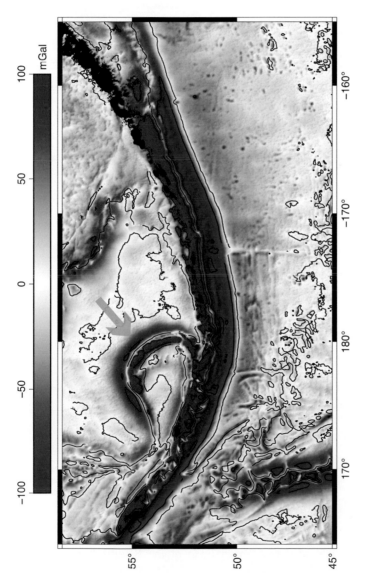

Figure 9.3 Gravity anomaly of the Aleutian Trench (100 mGal contour interval). The deepest part of the trench has large negative anomalies (−200 mGal). The outer rise south of the trench has smaller positive anomalies (20–40 mGal). Note the extinct trench marked by a light gray arrow has a large gravity anomaly demonstrating that trench flexures reflect bending of an elastic plate rather than a viscous plate since the viscous flexure would decay in a few million years.

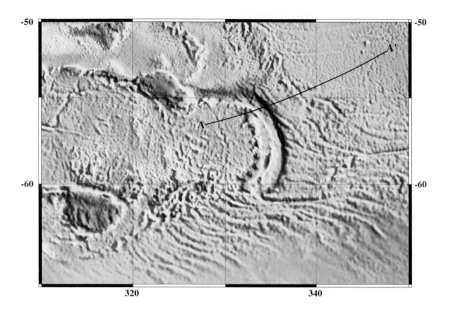

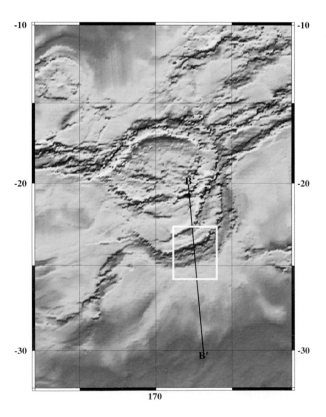

Figure 9.12 (upper) Shaded bathymetry map of the Sandwich trench, South Atlantic. (lower) Shaded topography of Latona Corona, Venus. The maps are plotted at the same horizontal and vertical scale (Sandwell and Schubert, 1992). Profiles A–A' and B–B' are profiles shown in Figure 9.13. The white box outlines the high resolution topography shown in Figure 9.14.

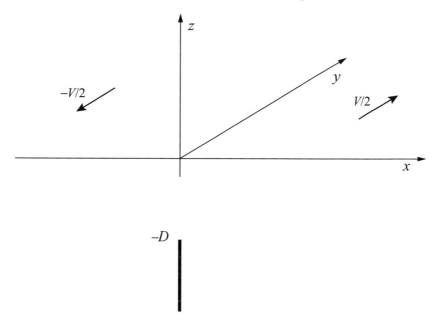

Figure 10.5

10.1.5 Vertical Integration of Line Source
to Create a Fault Plane

The final step in the development is to integrate the line-source screw dislocation over depth. We consider three cases:

1. Deep slip to represent interseismic deformation above a locked fault.
2. Shallow slip to represent shallow creep.
3. Shallow slip on a stress-free crack to represent an earthquake.

Case 1 First consider a fault that is free-slip between a depth $-D$ and infinity. This is the solution considered by Savage (1990). See Figure 10.5.

The integral of the line source Green's function is

$$v(x) = \frac{-A}{\pi \mu} \int_{-\infty}^{-D} \frac{x}{x^2 + z^2} \, dz. \qquad (10.19)$$

To integrate 10.19, make the following substitution.

$$\eta = -xz^{-1} \quad \text{so} \quad d\eta = xz^{-2} \, dz \qquad (10.20)$$

The integral becomes

$$v(x) = \frac{-A}{\pi\mu} \int_0^{x/D} \frac{1}{1+\eta^2} \, d\eta = \frac{-A}{\pi\mu} \tan^{-1}\left(\frac{x}{D}\right). \tag{10.21}$$

We know that $v(\pm\infty) = +V/2$, so $A = -V\mu$. Note that A has units of force per unit area times a velocity. This corresponds to a moment rate per area of fault. The familiar results for displacement and shear stress are

$$v(x) = \frac{V}{\pi}\tan^{-1}\frac{x}{D}$$

$$\tau_{xy} = \frac{\mu V}{\pi D} \frac{1}{1+\left(\frac{x}{D}\right)^2}. \tag{10.22}$$

Consider the extreme cases of a completely unlocked fault such that $D = 0$. The displacement field will be a step function and the stress will be everywhere zero except at the origin, where it will be infinite.

Case 2 Next consider a fault that is free-slip between the surface and a depth $-d$. In this case the integral is

$$v(x) = \frac{V}{\pi} \int_{-x/d}^{0} \frac{1}{1+\eta^2} \, d\eta = \frac{V}{\pi} \tan^{-1}\eta \Big|_{-x/d}^{\infty}. \tag{10.23}$$

There are two cases depending on whether x is positive or negative.

$$v(x) = \frac{V}{\pi}\left(\frac{\pi}{2} - \tan^{-1}\frac{x}{d}\right) \quad x > 0$$

$$v(x) = \frac{V}{\pi}\left(\frac{-\pi}{2} - \tan^{-1}\frac{x}{d}\right) \quad x < 0 \tag{10.24}$$

By combining these, the displacement and shear stress are

$$v(x) = V\left[H(x) - \tfrac{1}{2}\right] - \frac{V}{\pi}\tan^{-1}\frac{x}{d}$$

$$\tau_{xy} = \mu V\left[\delta(x) - \frac{1}{\pi d}\frac{1}{1+\left(\frac{x}{d}\right)^2}\right]. \tag{10.25}$$

If the fault is completely unlocked so that as d goes to infinity, the displacement becomes a step and the shear stress is infinite at the origin, in agreement with our concepts of a free-slipping fault.

Case 3 The third case considered also has shallow slip between depth $-d$ and the surface. However, in this case we consider a so-called *crack model*, where the slip versus depth function results in zero stress on the fault. This derivation will lead to the crack solution given in *Geodynamics*, equation (8.110). The Case 2 solution has uniform slip with depth. This leads to a stress singularity at the base of the fault. In contrast, the model in *Geodynamics* has a stress-free crack imbedded in a pre-stressed elastic half space. Using the Green's function developed above, it can be shown that the two solutions are in fundamental agreement. The only difference is related to the slip-versus-depth function.

From the dislocation theory developed in equation (10.19), the y-displacement as a function of distance from the fault is given by

$$v(x) = \frac{1}{\pi} \int_{-d}^{0} \frac{s(z)x}{x^2 + z^2} \, dz \tag{10.26}$$

where z is depth, x is distance from the fault, $s(z)$ is the slip versus depth, and $v(x)$ is the displacement. Now consider the two slip versus depth functions between the surface and $-d$.

$$s_1 = S$$
$$s_2 = S(1 - z^2/d^2)^{1/2} \tag{10.27}$$

The first slip function is constant with depth, while the second corresponds to the stress-free crack and has the form provided in *Geodynamics*, equation (8.93). Using the approach described above, the integral of the constant slip with depth s_1 is

$$v(x) = \frac{S}{\pi} \left(\frac{x}{|x|} \frac{\pi}{2} - \tan \frac{x}{d} \right). \tag{10.28}$$

The integral of the slip function for the crack model s_2 is given by

$$v(x) = \frac{S}{\pi} x \int_{-d}^{0} \frac{\left(1 - z^2/d^2\right)^{1/2}}{x^2 + z^2} \, dz = \frac{S}{\pi} x \int_{0}^{d} \frac{\left(1 - z^2/d^2\right)^{1/2}}{x^2 + z^2} \, dz. \tag{10.29}$$

Now we let $x' = x/d$ and $z' = z/d$, so the integral becomes

$$v(x') = \frac{S}{\pi} x' \int_{0}^{1} \frac{\left(1 - z'^2\right)^{1/2}}{x'^2 + z'^2} \, dz. \tag{10.30}$$

This integral can be performed in MATLAB using the following code with the symbolic toolbox.

```
clear
syms x positive
syms z
arg=sqrt(1-z*z)/(x*x+z*z);
int(arg,z,0,1)
```

Note that the integrand contains x'^2, so the results for positive and negative x' are identical. Therefore in the integrated result, the x' should be replaced by $|x'|$. The result is

$$v\left(x'\right) = \frac{S}{\pi}x'\frac{\pi}{2\,|x'|}\left[\left(1+x'^2\right)^{1/2} - |x'|\right].\tag{10.31}$$

Finally, substitute for x' and we arrive at

$$v\left(x\right) = \frac{x}{|x|}\frac{S}{2}\left[\left(1+\frac{x^2}{d^2}\right)^{1/2} - \frac{|x|}{d}\right].\tag{10.32}$$

This matches equation (8.110) in *Geodynamics*.

One can now make a direct comparison between the displacement versus distance for the two slip functions, to note their similarities and differences. However, realize that the arctangent slip function will have a larger seismic moment (i.e., slip integrated over depth) than the crack model slip function. The magnitude of the difference is found by integrating the slip versus depth for the two cases. For the arctangent function the integrated slip is simply Sd. For the crack model the integrated slip is

$$Sd\int_0^1 (1-z^2)^{1/2}\,\mathrm{d}z = Sd\frac{\pi}{4}.\tag{10.33}$$

Figure 10.6 compares the two displacement functions when the depth of faulting for the arctangent model is reduced by $\pi/4$, so the moments are matched; at this scale the plots are nearly identical. This illustrates the fact that measurements of displacement versus distance across a fault are not very sensitive to the shape of the slip versus depth function, although they do provide an important constraint on the overall seismic moment. In the next section, we highlight the issue that geodetic measurements of surface displacement are relatively insensitive to the shape of the slip (versus depth function), but provide a good estimate of the overall seismic moment.

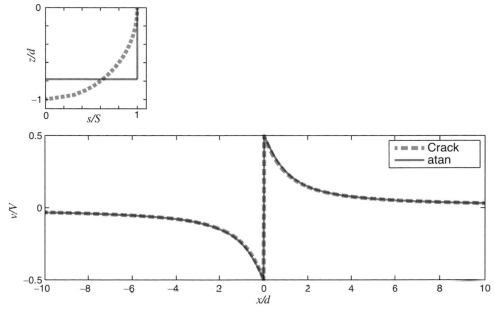

Figure 10.6

10.2 Geodetic Moment Accumulation Rate

The geodetic moment accumulation rate M per unit length of fault L is given by the well known formula

$$\frac{M}{L} = \mu S D \tag{10.34}$$

where D is the thickness of the locked zone and S is the slip deficit rate or backslip rate used in block models. This is the standard formula provided in all the seismology textbooks, although they usually consider the co-seismic moment release due to co-seismic slip. Here we are considering the gradual accumulation of geodetic moment during the interseismic period. These moments must balance over many earthquake cycles. In the general case where the slip rate s varies with depth z, the moment rate M is given by

$$\frac{M}{L} = \mu \int_{-D_m}^{0} s\,(z)\,\mathrm{d}z. \tag{10.35}$$

where D_m is the maximum slip depth. The objective of the following analysis is to show that the total moment rate per unit length of fault can be measured directly from geodetic data; no slip vs. depth model is needed. The only assumptions are that the strike-slip fault is 2-D and the Earth behaves as an elastic half space. From the dislocation theory developed in equation (10.26), the y-velocity as a function of distance from the fault is given by

$$v\,(x) = \frac{1}{\pi} \int_{-D_m}^{0} \frac{s(z)x}{x^2 + z^2}\, dz. \tag{10.36}$$

See also Figure 10.7.

Next we guess that the integral of the displacement rate times distance from the x-origin is a proxy for the moment accumulation rate. We call this proxy Q and later show how it is related to the moment rate M. We integrate to an upper limit W and then take the limit as $W \to \infty$.

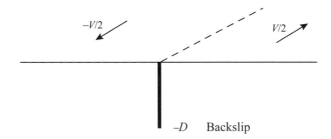

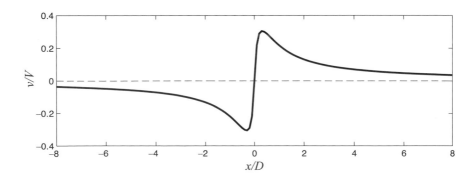

Figure 10.7 Schematic of surface velocity due to uniform backslip rate over a depth D.

$$Q = \lim_{W \to \infty} \left[\frac{1}{W} \int_0^W x v\,(x)\,dx \right] = \lim_{W \to \infty} \left[\frac{1}{\pi W} \int_0^W \int_{-D_m}^0 s(z) \frac{x^2}{x^2 + z^2}\,dz\,dx \right]$$

(10.37)

After rearranging the order of integration, one finds

$$Q = \frac{1}{\pi} \int_{-D_m}^0 s(z) \lim_{W \to \infty} \left(\frac{1}{W} \int_0^W \frac{x^2}{x^2 + z^2}\,dx \right) dz.$$

(10.38)

The integral over x can be done analytically.

$$\frac{1}{W} \int_0^W \frac{x^2}{x^2 + z^2}\,dx = \frac{x}{W} - \frac{z}{W} \tan^{-1} \frac{x}{z} \Big|_o^W = 1 - \frac{z}{W} \tan^{-1} \frac{W}{z}$$

(10.39)

In the limit as $W \to \infty$, the second term on the right side is zero, because z has an upper bound of D_m; so the total integral is simply 1. Overall, we find this proxy is

$$Q = \frac{1}{\pi} \int_{-D_m}^0 s(z)\,dz.$$

(10.40)

Comparing equation (10.37) with equation (10.40), it is clear that the geodetic moment can be directly related to the integral of the displacement times the distance from the origin. Note we have extended the integral to both sides of the fault to enable the use of geodetic measurements on both sides.

$$\frac{M}{L} = \lim_{W \to \infty} \left[\frac{\mu \pi}{W} \int_{-W}^W x v\,(x)\,dx \right]$$

(10.41)

As a check, we can insert equation (10.36) into equation (10.41) and make sure we arrive at equation (10.35).

$$\frac{M}{L} = \lim_{W \to \infty} \left[\frac{\mu \pi}{W} \int_{-W}^W x \left(\frac{1}{\pi} \int_{-D_m}^0 \frac{s\,(z)\,x}{x^2 + z^2}\,dz \right) dx \right]$$

(10.42)

$$= \mu \int_{-D_m}^0 s\,(z) \left(\lim_{W \to \infty} \frac{1}{W} \int_{-W}^W \frac{x^2}{x^2 + z^2}\,dx \right) dz$$

We perform the integral over x first and multiply by 2 after changing the limits, because the integrand is symmetric about $x = 0$.

$$2 \int_0^W \frac{x^2}{x^2 + z^2} \, dx = 2 \left(x - z \tan^{-1} \frac{x}{z} \right) \bigg|_0^W \tag{10.43}$$

In the limit as $W \to \infty$, the final result is

$$\lim_{W \to \infty} \frac{2}{W} \left(W - z \tan^{-1} \frac{W}{z} \right) = 2, \quad \text{for } z_{max} = D_m \ll W. \tag{10.44}$$

The moment accumulation rate is

$$\frac{M}{L} = 2\mu \int_{-D_m}^0 s\,(z)\, dz. \tag{10.45}$$

This agrees with our original estimate of moment, except for a factor of 2. The corrected formula for the moment accumulation rate is

$$\frac{M}{L} = \lim_{W \to \infty} \frac{\mu\pi}{2W} \int_{-W}^W x v\,(x)\, dx. \tag{10.46}$$

The main utility of this formula is to demonstrate that geodetic measurements of y-displacement rate across an infinitely long strike-slip fault provides a direct estimate of the geodetic moment rate. It is unnecessary to attempt the unstable inverse problem to calculate slip versus depth and then integrate this function.

10.3 Inclined Fault Plane

Now consider a model where the fault plane is not perpendicular to the free surface of the Earth as shown in Figure 10.8. The angle α between the vertical and the fault plane will introduce an asymmetry in the model. To develop this solution, we'll start with the surface displacement due to a screw dislocation. We'll integrate over depth and rotate from the inclined frame into the horizontal frame. Finally, we'll introduce the image source to reconcile the free surface boundary condition.

From equation (10.17), we have

$$v(x', z') = \frac{-A}{2\pi\mu} \frac{x'}{\left[x'^2 + (z' + a)^2 \right]}. \tag{10.47}$$

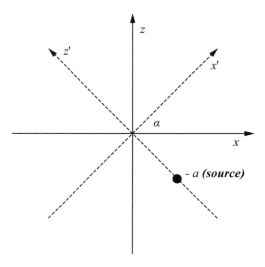

Figure 10.8

The rotation from the x, z frame to the x', z' frame is

$$x' = x \cos \alpha + z \sin \alpha$$
$$z' = -x \sin \alpha + z \cos \alpha. \tag{10.48}$$

Also note that $D = D' \cos \alpha$. As before, consider free slip between a depth of $-D'$ and minus infinity.

$$v(x',z') = \frac{V}{2\pi} \int_{-\infty}^{-D'} \frac{x'}{x^2 + (z' + a')^2} \, da' \tag{10.49}$$

Let $\eta = z' + a'$, so $d\eta = da'$.

$$v(x',z') = \frac{V}{2\pi} \int_{-\infty}^{-z'-D'} \frac{x'}{x'^2 + \eta^2} \, d\eta \tag{10.50}$$

We have performed this integration before (equations (10.19)–(10.20)), so it is not repeated here. The result is

$$v(x',z') = \frac{V}{2\pi} \tan^{-1} \left(\frac{x'}{D' + z'} \right). \tag{10.51}$$

To match the surface boundary condition, we introduce an image source extending from $+D'$ to infinity, but along an image fault inclined at an angle of $-\alpha$ with respect to the vertical. The displacement from the image is:

$$v_{\text{image}}(x'', z'') = \frac{V}{2\pi} \tan^{-1}\left(\frac{x''}{D'' - z''}\right). \tag{10.52}$$

Finally, combining the source and the image and substituting x and z we find

$$v(x, z) = \frac{V}{2\pi}\left\{\tan^{-1}\left(\frac{x\cos\alpha + z\sin\alpha}{D' - x\sin\alpha + z\cos\alpha}\right) + \tan^{-1}\left(\frac{x\cos\alpha - z\sin\alpha}{D' - x\sin\alpha - z\cos\alpha}\right)\right\}. \tag{10.53}$$

Now calculate the displacement at $z = 0$ and substitute $D' = D/\cos\alpha$.

$$v(x) = \frac{V}{\pi}\tan^{-1}\left(\frac{x\cos^2\alpha}{D - x\sin\alpha\,\cos\alpha}\right) \tag{10.54}$$

This dipping fault case has two differences from the vertical strike-slip fault case. First, the displacement pattern is shifted along the x-axis by an amount $D\tan\alpha$. Therefore, one can identify a dipping fault by recognizing that the position of the fault based on geodetic measurements is shifted from the position of the fault trace based on field geology.

The second difference is that the solution given in equation (10.54) does not match the far-field boundary conditions of $\pm V/2$. The hanging wall has more displacement than the foot wall. In the extreme case of a near horizontal fault plane, the hanging wall has the full displacement $+V$, while the foot wall has none. This is to be expected, because the model is driven by a force couple. One can "correct" this asymmetry by subtracting a constant α from the arctangent in equation (10.54). It is left as an exercise for the reader to show the final solution is

$$v(x) = \frac{V}{\pi}\left[\tan^{-1}\left(\frac{x\cos^2\alpha}{D - x\sin\alpha\,\cos\alpha}\right) - \alpha\right]. \tag{10.55}$$

We see that for $\alpha = 0$, this matches the previous solution, equation (10.22). Also, we can superimpose several of these solutions to simulate any combination of shallow and deep slip. The stress is the shear modulus times the x-derivative of the displacement. After a little algebra one finds

$$\tau_{xy} = \frac{\mu V}{\pi D_\alpha}\left[1 + \left(\frac{x\cos^2\alpha}{D_\alpha}\right)^2\right]^{-1} \cdot \left[\cos^2\alpha + \frac{x\cos^3\alpha\,\sin\alpha}{D_\alpha}\right] \tag{10.56}$$

where $D_\alpha = D - x\sin\alpha\,\cos\alpha$.

10.4 MATLAB **Examples**

The first example is a MATLAB program to calculate the strain and displacement fields due to a vertical strike-slip fault with free-slip on both shallow and deep fault planes.

```
%
%  program to generate displacement and strain for a screw
%  dislocation. fault slip occurs both shallow and deep.
%
clear
clf
hold off
%
V=-.01;
D=12000.;
d=800.;
d0=200;
x = -40000:8:40000;
xp = x/1000.;
%
%  this first model has shallow creep between depths of d0 and d
%
v1 = (V/pi)*(atan(x/d0)-atan(x/d));
dv1 = (V/(pi*d0))*1./(1.+(x/d0).^{2}) - (V/(pi*d))*1./(1.+(x/d).^{2});
%
%  this second model has free-slip for depths greater than D.
%
v2 = (V/pi)*atan(x/D);
dv2 = (V/(pi*D))*1./(1.+(x/D).^{2});
%
subplot(2,1,2);plot(xp,(v1+v2)*1000,xp,v2*1000,':');
        xlabel('distance (km)');
        ylabel('displacement (mm/a)')
subplot(2,1,1);plot(xp,1.e6*(dv1+dv2),xp,1.e6*dv2,':');
        ylabel('strain (microradian/a)'); axis([-40,40,-3,1])
```

See Figure 10.9

The second example is a MATLAB program to illustrate the effect of fault dip that simply shifts the arctangent function by an amount $D\tan\alpha$. In this example, the shift is 6.9 km.

```
%
%  Compute the displacement due to a dipping fault using equation (34).
%  Note the function atan2() must be used.
%
V=-10;
alph=30*pi/180.;
D=12;
x=-40:40;
```

```
%
%
cosa=cos(alph);
sina=sin(alph);
num=x.*cosa*cosa;
dem=D-x.*sina*cosa;
vel0=V*atan2(x,D)/pi;
vel1=V*(atan2(num,dem)/pi-alph/pi);
subplot(2,1,1);plot(x,vel0,x,vel1,'--');
xlabel('distance (km)');ylabel('displacement (mm/a)');
title('dipping 30 degrees in positive x-direction')
grid
%
```

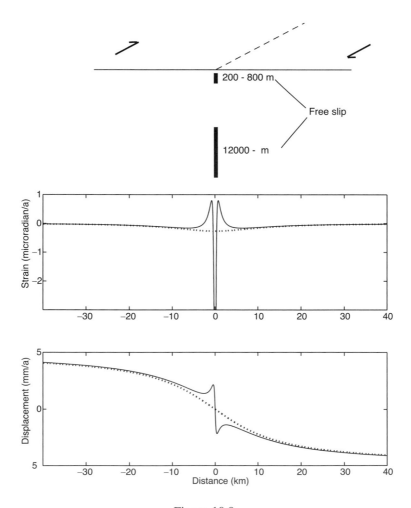

Figure 10.9

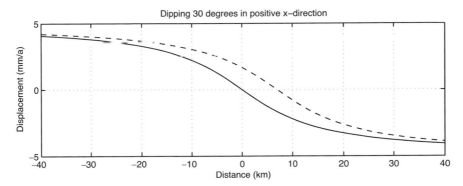

Figure 10.10

See Figure 10.10.

10.5 Exercises: Response of an Elastic Half Space to a 3-D Vector Body Force

This exercise illustrates the power and flexibility of the Fourier transform approach to solving complicated linear partial differential equations. We wish to calculate the displacement vector $\mathbf{u}\,(x, y, z)$ on the surface or inside of the Earth due to a vector body force at depth. This solution can be used to construct a variety of models such as fault slip of arbitrary complexity (e.g. Smith and Sandwell, 2003). We start with the equations that relate stress to body forces, stress to strain, and strain to displacement.

$$\sigma_{ij,j} = -b_i \tag{10.57}$$

$$\sigma_{ij} = \delta_{ij}\lambda\varepsilon_{kk} + 2\mu\varepsilon_{ij} \tag{10.58}$$

$$\varepsilon_{ij} = \frac{1}{2}\left(u_{i,j} + u_{j,i}\right) \tag{10.59}$$

Substitute equation 10.59 into equation 10.58.

$$\sigma_{ij} = \delta_{ij}\lambda u_{k,k} + \mu\left(u_{i,j} + u_{j,i}\right) \tag{10.60}$$

Substitute equation 10.60 into equation 10.57

$$\delta_{ij}\lambda u_{k,kj} + \mu\left(u_{i,jj} + u_{j,ij}\right) = -b_i \tag{10.61}$$

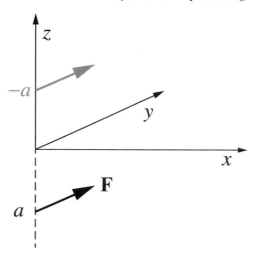

Figure 10.11 Coordinate system with point vector body forces at a and $-a$.

and rewrite as

$$(\lambda + \mu)\, u_{k,ki} + \mu u_{i,kk} = -b_i. \tag{10.62}$$

Equation 10.62 can be written as three partial differential equations.

$$\mu \nabla^2 u + (\lambda + \mu) \left[\frac{\partial^2 u}{\partial x^2} + \frac{\partial^2 v}{\partial y \partial x} + \frac{\partial^2 w}{\partial z \partial x} \right] = -b_x$$

$$\mu \nabla^2 v + (\lambda + \mu) \left[\frac{\partial^2 u}{\partial x \partial y} + \frac{\partial^2 v}{\partial y^2} + \frac{\partial^2 w}{\partial z \partial y} \right] = -b_y \tag{10.63}$$

$$\mu \nabla^2 w + (\lambda + \mu) \left[\frac{\partial^2 u}{\partial x \partial z} + \frac{\partial^2 v}{\partial y \partial z} + \frac{\partial^2 w}{\partial z^2} \right] = -b_z$$

Next introduce a vector body force at $x = y = 0$, $z = a$. To satisfy the boundary condition of zero shear traction at the surface, an image source is also applied at $x = y = 0$, $z = -a$ (Figure 10.11). Equation 10.64 describes this point body force at the source and image locations, where $\mathbf{F} = (F_x, F_y, F_z)$ is a vector force with units of force.

$$\mathbf{b}\,(x, y, z) = \mathbf{F}\delta\,(x)\,\delta\,(y)\,\delta\,(z - a) + \mathbf{F}\delta\,(x)\,\delta\,(y)\,\delta\,(z + a) \tag{10.64}$$

Exercise 10.1 It is left as an exercise to take the 3-D Fourier transform of equations 10.63 and 10.64 to reduce the partial differential equations to a set of linear algebraic equations. Several of the properties of Fourier transforms, given in Chapter 2, will be helpful. The result is:

$$(\lambda + \mu) \begin{bmatrix} k_x^2 + \frac{\mu \mathbf{k}^2}{(\lambda + \mu)} & k_y k_x & k_z k_x \\ k_x k_y & k_y^2 + \frac{\mu \mathbf{k}^2}{(\lambda + \mu)} & k_z k_y \\ k_x k_z & k_y k_z & k_z^2 + \frac{\mu \mathbf{k}^2}{(\lambda + \mu)} \end{bmatrix} \begin{bmatrix} U(\mathbf{k}) \\ V(\mathbf{k}) \\ W(\mathbf{k}) \end{bmatrix}$$

$$= \frac{e^{-i2\pi k_z a} + e^{i2\pi k_z a}}{4\pi^2} \begin{bmatrix} F_x \\ F_y \\ F_z \end{bmatrix} \tag{10.65}$$

where $\mathbf{k} = (k_x, k_y, k_z)$ and $\mathbf{k}^2 = \mathbf{k} \cdot \mathbf{k}$.

Exercise 10.2 Invert the linear system of equations to isolate the 3-D displacement vector solution for $U(\mathbf{k})$, $V(\mathbf{k})$, and $W(\mathbf{k})$. This can be done using the symbolic algebra capabilities of MATLAB or another computer algebra package.

$$\begin{bmatrix} U(\mathbf{k}) \\ V(\mathbf{k}) \\ W(\mathbf{k}) \end{bmatrix} = \frac{(\lambda + \mu)\left(e^{-i2\pi k_z a} + e^{i2\pi k_z a}\right)}{\mu(\lambda + 2\mu)4\pi^2 \mathbf{k}^4}$$

$$\times \begin{bmatrix} \left(k_y^2 + k_z^2\right) + \frac{\mu \mathbf{k}^2}{(\lambda + \mu)} & -k_y k_x & -k_z k_x \\ -k_x k_y & \left(k_x^2 + k_z^2\right) + \frac{\mu \mathbf{k}^2}{(\lambda + \mu)} & -k_z k_y \\ -k_x k_z & -k_y k_z & \left(k_x^2 + k_y^2\right) + \frac{\mu \mathbf{k}^2}{(\lambda + \mu)} \end{bmatrix} \begin{bmatrix} F_x \\ F_y \\ F_z \end{bmatrix}$$

$$\tag{10.66}$$

One can check the inversion using the following MATLAB symbolic code.

```
%
%   MATLAB routine to check the solution in the kx, ky, kz domain.
%
    pi = sym('pi');
    kx=sym('kx');
    ky=sym('ky');
    kz=sym('kz');
% elastic constants
    la=sym('la');
    mu=sym('mu');
    lam=la+mu;
% combinations of wavenumbers
    c=sym('c');
    c=(kx*kx+ky*ky+kz*kz);
% forward matrix
    A=[kx*kx+mu*c/lam,ky*kx,   kz*kx;
       kx*ky,ky*ky+mu*c/lam,kz*ky;
       kx*kz,ky*kz,kz*kz+mu*c/lam];
% solution in Fourier domain, inverse matrix
    B=[c*mu/lam+ky*ky+kz*kz,-kx*ky,-kx*kz;
       -kx*ky,c*mu/lam+kx*kx+kz*kz,-kz*ky;
       -kx*kz,-ky*kz,   c*mu/lam+kx*kx+ky*ky];
```

```
% normalize
   A=A*lam;
   B=B*lam/(mu*(la+2*mu)*c*c);
% multiply to get the identity matrix
   C=B*A;
   simplify(C)
ans =

[ 1, 0, 0]
[ 0, 1, 0]
[ 0, 0, 1]
```

The next step is to perform the inverse Fourier transform with respect to k_z of each of the three terms for each of the three displacement components for both the source and the image. There are 12 different integrations to be performed although they are all very similar. For example, the following is the inverse transform for the $V(\mathbf{k})$ displacement driven by the F_x component of body force (source not image).

$$[V(\mathbf{k}_h, z) = -F_x \frac{(\lambda + u)}{\mu(\lambda + 2u)} \frac{k_x k_y}{4\pi^2} \int_{-\infty}^{\infty} \frac{e^{i2\pi k_z(z-a)}}{(k_z + i|\mathbf{k}_h|)^2 (k_z - i|\mathbf{k}_h|)^2} \, dk_z \quad (10.67)$$

where $\mathbf{k}_h = (k_x, k_y)$ is the horizontal wavenumber and $|\mathbf{k}_h| = (k_x^2 + k_y^2)^{1/2}$. The denominator has four poles in the complex plane, two at $-i|\mathbf{k}_h|$ and two at $i|\mathbf{k}_h|$.

> **Exercise 10.3** Use the Cauchy residue theorem, for the case of repeated poles, and the boundary condition that the displacement must vanish as $z \to \infty$ to derive the $V(\mathbf{k})$ displacement.

$$V(\mathbf{k}_h, z) = -F_x \frac{(\lambda + u)}{\mu(\lambda + 2u)} \frac{k_x k_y}{2\pi} \left[\frac{1 + 2\pi |\mathbf{k}_h|}{4|\mathbf{k}_h|^3} \right] e^{-2\pi |\mathbf{k}_h|(z-a)} \quad (10.68)$$

There are a number of additional steps needed to develop a full algorithm for computing displacements due to double-couple fault sources. These are published in Smith and Sandwell (2003). The astute reader will notice that while the introduction of an image body force achieves the zero shear stress boundary condition, the normal stress at the surface is not zero. This boundary condition can be corrected by applying an equal but opposite vertical stress following Steketee (1958). Also, for purely vertical faults, the body force can be analytically integrated over depth. The final step is a numerical implementation for a complicated fault model. This involves making 2-D grids of each of the three components of the body force. The forces in these grids can be arranged to create single or double couples in an arbitrarily complex pattern. For a more complete description, see Smith and Sandwell (2003). These three force grids are Fourier transformed and multiplied by the elements of the Earth response (e.g., equation (10.68)), The results are summed and inverse transformed to calculate the vector displacement as well as all the stress

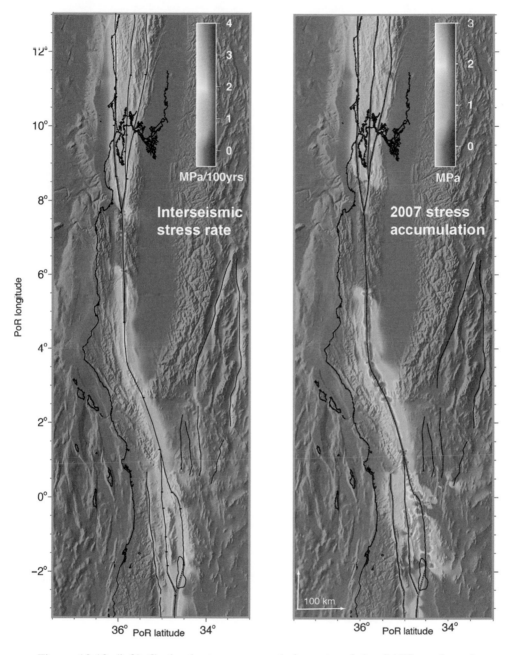

Figure 10.12 (left) Coulomb stress accumulation rate of the SAFS, evaluated at half of the locking depth projected into pole of rotation coordinate system (Wdowinski et al., 2007). (right) Present-day (calendar year 2007) Coulomb stress accumulation based on stress accumulation and contributions from 75 historical and prehistorical earthquake ruptures. Black lines are faults where force double couples are applied and imbedded in a 1 km by 1 km grid. Modified from Smith-Konter and Sandwell (2009).

components through differentiation. The main advantages of this approach are computational efficiency and the ability to construct very complex fault systems. An example of the Coulomb stress accumulation rate on the San Andreas Fault system is shown in Figure 10.12. Note this analysis was extended to include a uniform-thickness elastic plate over a viscoelastic half space (Smith and Sandwell, 2004). The computer code for creating these 3-D, time-dependent fault models can be found at: `github.com/dsandwell/fftfault`.

11

Heat Flow Paradox

11.1 Heat Flow Paradox

The heat flow paradox relates the expected frictional heating on a strike-slip fault, such as the San Andreas Fault, to the measurements of surface heat flow above the fault (e.g. Lachenbruch and Sass, 1980). A straightforward calculation, using a reasonable coefficient of friction for the fault, predicts measurably high heat flow above the fault that is not observed. Is the lack of a heat flow anomaly evidence that faults are very weak or is the heat flow anomaly erased by hydrothermal circulation in the crust? In this chapter we develop a simple heat conduction model, following Lachenbruch and Sass (1980), to explore this paradox.

The seismogenic zone extends from the surface to a depth of about 12 km. According to Byerlee's law (Byerlee, 1978), the shear stress on the fault should be some fraction of the lithostatic stress.

$$\tau(z) = f \left(\rho_c - \rho_w \right) g z \tag{11.1}$$

f	static coefficient of friction	~ 0.60
ρ_c	crustal density	2600 kg m^{-3}
ρ_w	water density	1000 kg m^{-3}
g	acceleration of gravity	9.8 m s^{-2}
D	depth of seismogenic zone	12 km

This assumes that water percolates to 12 km depths to lower friction on the fault. We can compute the average shear stress on the fault.

$$\bar{\tau} = \frac{1}{D} \int_o^D f \left(\rho_c - \rho_w \right) g z \, dz = \tfrac{1}{2} f \left(\rho_c - \rho_w \right) g D = 56 \text{ MPa} \tag{11.2}$$

169

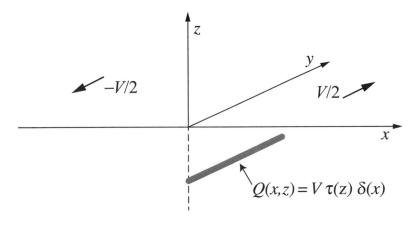

Figure 11.1

The observed stress drop during an earthquake ranges from 0.1 to 10 MPa, with a typical value of 5 MPa, which is about 10 times smaller than the average stress from Byerlee's law (Byerlee, 1978). This implies that only a fraction of the total stress is released during an earthquake. The average stress during the earthquake times the earthquake displacement produces energy both as seismic radiation (small fraction) and as heat (large fraction). If this heat energy is averaged over many earthquake cycles, then this average heat/area generated on the fault plane will appear as a heat flow anomaly on the surface having a similar heat/area as along the fault.

To calculate this heat anomaly for a variety of frictional heating models, first consider a line source of heat. (See Figure 11.1.) The differential equation and boundary conditions for a unit-amplitude, line source at depth $-a$ is

$$\nabla^2 T = \frac{1}{k} Q(x,z) = \frac{1}{k} \delta(x)\,\delta(z+a) \tag{11.3}$$

$$T(x,0) = 0$$

$$\lim_{|z|\to\infty} T(x,z) = 0$$

$$\lim_{|x|\to\infty} T(x,z) = 0$$

where T is the temperature anomaly, k is the thermal conductivity (3.3 W m^{-1} K^{-1}) and Q is the heat generation in W m^{-3}. Note this is the same differential equation as equation (10.5). The only difference is the surface boundary condition. The strike-slip fault problem has vanishing shear stress at the surface (i.e., the vertical derivative of displacement v is zero), so we introduced a positive image source to force the displacement field to be symmetric about $z = 0$. In this heat flow case, we have vanishing temperature anomaly at the surface, so we introduce a negative line heat source at $z = a$ to form an anti-symmetric temperature function The solution to the full space problem is identical to equation (10.14). Note this problem was also solved in Section 2.6.

$$T(x,z) = \frac{-1}{2\pi k} \ln\left[x^2 + (z+a)^2\right]^{1/2} \qquad (11.4)$$

After including the image source, the result is

$$T(x,z) = \frac{-1}{2\pi k}\left\{\ln\left[x^2 + (z+a)^2\right]^{1/2} - \ln\left[x^2 + (z-a)^2\right]^{1/2}\right\}. \qquad (11.5)$$

The quantity of interest is the surface heat flow versus distance from the fault.

$$q(x,z) = -k\frac{\delta T}{\delta z} = \frac{1}{2\pi}\frac{\delta}{\delta z}\left\{\ln\left[x^2 + (z+a)^2\right]^{1/2} - \ln\left[x^2 + (z-a)^2\right]^{1/2}\right\} \qquad (11.6)$$

After a little algebra, one arrives at the heat flow.

$$q(x,z) = \frac{1}{2\pi}\left\{\frac{(z+a)}{x^2 + (z+a)^2} - \frac{(z-a)}{x^2 + (z-a)^2}\right\} \qquad (11.7)$$

Thus, the surface heat flow for a line source of unit strength at depth a is

$$q(x) = \frac{1}{\pi}\frac{a}{x^2 + a^2}. \qquad (11.8)$$

For an arbitrary shear stress distribution with depth $\tau(z)$, the surface heat flow is

$$q(x) = \frac{V}{\pi}\int_{-\infty}^{0}\frac{z\,\tau(z)}{x^2 + z^2}\,dz. \qquad (11.9)$$

Now let's assume that the stress follows equation (11.1), Byerlee's law (i.e., high stress and high heat flow). Also allow hydrothermal circulation to extend from the surface to some depth d, which effectively removes all the heat produced between the surface and that depth. The integration is

$$q(x) = \frac{f(\rho_c - \rho_w)\,gV}{\pi}\int_{-D}^{-d}\frac{z^2}{x^2 + z^2}\,dz. \qquad (11.10)$$

This integral is done with help from the table of integrals.

$$\int\frac{x^2}{a + bx^2}\,dx = \frac{x}{b} - \frac{a}{b}\int\frac{1}{a + bx^2}\,dx \qquad (11.11)$$

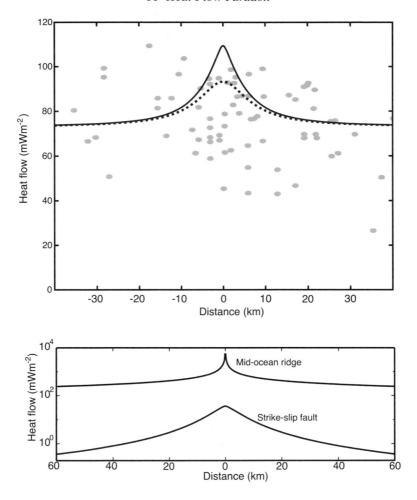

Figure 11.2 (upper) Heat flow measurements across the San Andreas Fault (Lachenbruch and Sass, 1980) compared with model predictions. The solid curve is heat flow with shallow hydrothermal circulation while the dashed curve has hydrothermal heat removal to 5 km depth. (lower) Comparison of the heat flow generated on a strike-slip fault to heat flow at a mid ocean ridge showing it is at least 100 times smaller.

After some algebra, one arrives at the following analytic formula for the heat flow.

$$q(x) = \frac{f(\rho_c - \rho_w)gV}{\pi} \left\{ (D - d) + \left(x \tan^{-1} \frac{d}{x} - x \tan^{-1} \frac{D}{x} \right) \right\} \qquad (11.12)$$

It is interesting to compare this heat flux to the heat flux at a mid-ocean ridge for the same total opening rate V (see Figure 11.2). The formula is

$$q(x) = k(T_m - T_o)(2\pi\kappa x/V)^{-1/2}. \qquad (11.13)$$

11.1.1 MATLAB Example

The following MATLAB program simulates a high-stress fault (i.e., Byerlee's law) extending to a depth of 12 km and sliding at a rate of 30 mm/yr. Two cases are considered; the first case (the solid curve in Figure 11.2) has hydrothermal heat removal extending to a depth of 1 km, while the second case (the dotted curve in the same figure) has heat removal to a depth of 5 km. These models are compared with the heat flow measurements across the San Andreas Fault (Lachenbruch and Sass, 1980). It is clear that the shallow heat removal model is inconsistent with the data. However, the deep heat removal model is not precluded by the observations, especially if the background level of the model heat flow is allowed to vary from the spatial average. One argument against hydrothermal removal of heat is the absence of hot springs along the fault with sufficient vigor to remove this heat. Hydrothermal circulation is the dominant heat removal mechanism at the mid-ocean ridges and hydrothermal vents are common. However, as shown in Figure 11.2, the heat generation along a strike-slip fault is two to three orders of magnitude less than a mid-ocean ridge, so it is not clear that the same mechanism should operate at a fault. Even if heat loss is concentrated in small areas, it may be difficult to detect at the surface.

```
%
% program to calculate the surface heat flux due to frictional
% heating on a strike-slip fault
D=12;
d1=1; d5=5;
rc=2600; rw=1000; g=9.8;
V=.03/3.15e7; f=.60;
q0=1.e6*f*(rc-rw)*g*V/pi;
%
% calculate the heat flow for the two models of shallow
% and deep heat removal
x=-60:.1:60;
q1=q0*((D-d1)+x.*atan(d1./x)-x.*atan(D./x));
q5=q0*((D-d5)+x.*atan(d5./x)-x.*atan(D./x));
% plot the results
plot(x,q1+73,x,q5+73,':');
xlabel('distance (km)');
ylabel('heat flow (mWm-2)')
axis([-40,40,0,120]);
```

11.2 Seismic Moment Paradox

The *seismic moment paradox* described next is really part of the heat-flow paradox, except that it is expressed in a different way. As discussed in Chapter 7, and in Brace and Kohlstedt (1980), measurements of stress difference in the uppermost crust to depths of several kilometers are consistent with a yield strength model following Byerlee's law. The static frictional resistance to sliding is a fraction f (0.6) of the overburden pressure of $\Delta\rho g z$. This leads to a fault strength of 100 MPa at a depth of only 10 km. We also found that these high stresses are required to support the 5000 m elevation of Tibet relative to India. This isostatic model is the minimum stress needed to support topography, so it is clear that high stresses exist at shallow depths in the crust. Before comparing the tectonic stress with the earthquake stress drop, it is useful to compare moment release from a large earthquake with the theoretical maximum moment that a Byerlee-strength fault can sustain without slipping.

11.2.1 Seismic Moment Released During an Earthquake

The moment released during an earthquake is given by

$$M_s = \mu L D \Delta y. \tag{11.14}$$

See Figure 11.3.

We will use the 1992 Landers M7.2 earthquake as an example. The parameters given in the Table 11.1 result in a moment of 9.8×10^{19}N m or a moment per unit fault length of 1.4×10^{15}N.

The Landers earthquake moment matches the published value and the recurrence interval of $\Delta y / V = 3000$ years seems reasonable for a fault out in the Mojave desert away from the San Andreas Fault. So everything seems consistent. Next, let's assume that the stress on the fault, as a function of depth, matches Byerlee's

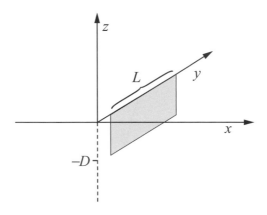

Figure 11.3

Table 11.1.

f	static coefficient of friction	~0.60
ρ_c	crustal density	2600 kg m^{-3}
ρ_w	water density	1000 kg m^{-3}
g	acceleration of gravity	9.8 m s^{-2}
μ	shear modulus	2.6 × 10^{10} Pa
L	length of rupture	70 km
D	depth of rupture	12 km
Δy	rupture offset	4.5 m
V	plate velocity	0.0015 m/yr
t	earthquake recurrence interval	

law for the case of hydrostatic pore pressure. We'll compare this saturation moment and recurrence interval with the observations from earthquakes.

11.2.2 Tectonic Saturation Moment

Assume that the simple half-space solution (developed above) provides the stress and strain field for a fault locked from the surface to a depth D. Further, assume that the maximum stress that can be maintained on a fault is given by Byerlee's law,

$$\tau(z) = f\left[-\left(\rho_c - \rho_w\right)gz + \tau_n\right] \tag{11.15}$$

where τ_n is the additional tectonic normal stress applied to the fault plane. The tectonic moment per unit length is given by

$$M_T/L = \Delta y \int_{-D}^{0} \mu(z)\,\mathrm{d}z. \tag{11.16}$$

What is $\mu(z)$? This is the effective shear modulus needed to keep the stress below the upper bound provided by Byerlee's law, so

$$\mu(z) = \frac{\tau(z)}{\varepsilon(z)} \quad \text{where} \quad \varepsilon(z) = \frac{\partial v(x,z)}{\partial x}. \tag{11.17}$$

Now assume that $v(x,z)$ is provided by the interseismic strain solution developed in the previous chapter (equation (10.18)). It is left as an exercise to finish the problem. You will find that $\varepsilon(z)$ is proportional to Δy, so this factor cancels in equation (11.16). The final result is

$$\frac{M_T}{L} = f\pi D^2 \left[\frac{\left(\rho_c - \rho_w\right)gD}{4} + \frac{2}{3}\tau_n\right]. \tag{11.18}$$

Using the values in Table 11.1 and for zero normal stress, we find the saturation moment per unit length is 1.3×10^{16} N. Again, this is 10 times larger than the moment per length for the Landers earthquake. Given the fault parameters above, this moment implies a potential seismic offset of 45 m and a recurrence time of 30,000 years—a giant earthquake indeed!

There are only two ways to understand this dilemma:

1. Faults are somehow lubricated ($f \sim 0.05$), so the average stress on the fault is 10–20 times smaller than predicted by Byerlee's law. In this case, one has the difficulty of maintaining the elevation of the topography in California. For example, San Jacinto Mountain, which is less than 25 km from the San Andreas Fault, has a relief of about 3000 m, which implies stresses of 80 MPa (16 times the stress drop in an earthquake).

2. Faults are strong as predicted by Byerlee's Law. In this case, faults are always very close to failure and each earthquake relieves only a small fraction ($\sim 10\%$) of the tectonic stress. As we saw in the last section, this model implies a large amount of energy dissipation along the fault; friction from both aseismic creep and seismic rupture will generate heat. It has been proposed that perhaps during the earthquake, the coefficient of friction drops from 0.60 to, say, 0.05, to temporarily disable the heat generation. However, it seems that such a slippery fault would release all of the elastic energy during an earthquake (~ 45 m of offset). Another possibility is that heat is generated, but a large fraction of the heat is advected to the surface by circulation of water in the upper couple of kilometers of crust. The unfortunate implication of this high-stress model is that since faults are always close to failure, it will be almost impossible to predict earthquakes.

11.3 Exercises

Exercise 11.1

(a) Provide an approximate formula for the magnitude of the shear stress that is needed to induce slip on a dry fault at 10 km depth in continental crust (density 2800 kg m^{-3}). Which parameter is least well known and what is a possible range for this parameter?

(b) Suppose the crust is saturated with water to 10 km depth. How does this change the stress magnitude?

Exercise 11.2 Derive equation 11.18.

12

The Gravity Field of the Earth, Part 1

12.1 Introduction

Chapters 12 through 15 cover *physical geodesy*, the shape of the Earth and its gravity field. This is electrostatic theory applied to the Earth. Unlike electrostatics, geodesy is a nightmare of unusual equations, unusual notation, and confusing conventions. Here we attempt to simplify and condense *physical geodesy* by focusing on concepts that are central to the field of geodynamics and tectonics. Chapter 5 of *Geodynamics* (Turcotte and Schubert, 2014) covers much of this topic but at a lower mathematical level. For a much more complete discussion of potential theory applied to the Earth, we recommend the excellent book by Blakely (1995). Following Blakely (1995) and Parker (1973), most of the Cartesian calculations are performed in the Fourier transform domain which greatly simplifies operations such as upward and downward continuation as well as modeling of complex density interfaces. Satellite radar altimetry has revolutionized our understanding of the gravity field, tectonics, and topography of the oceans so we focus on methods of processing and gridding these data in Chapter 15.

The things that make physical geodesy messy include:

- Earth rotation;
- latitude is measured from the equator instead of the pole;
- latitude is not the angle from the equator, but is referred to the ellipsoid;
- elevation is measured from a theoretical surface called the geoid;
- spherical harmonics are defined differently from standard usage;
- anomalies are defined with respect to an ellipsoid having parameters that are constantly being updated;
- there are many types of anomalies related to various derivatives of the potential; and
- MKS units are not commonly used in the literature.

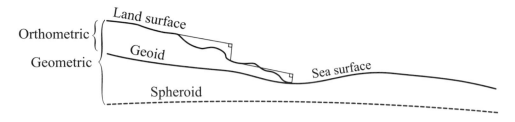

Figure 12.1

In the next couple of chapters, we will try to present this material with as much simplification as possible. Part of the reason for the mess is that prior to the launch of artificial satellites, measurements of elevation and gravitational acceleration were all done on the surface of the Earth (land or sea). Since the shape of the Earth is linked to variations in gravitational potential, measurements of acceleration were linked to position measurements both physically and in the mathematics. Satellite measurements are made in space well above the complications of the surface of the Earth, so most of these problems disappear. Here are the two most important issues related to old-style geodesy.

12.1.1 Elevation

Prior to satellites and the global positioning system (GPS), elevation was measured with respect to sea level which is approximately an equipotential surface called the *geoid*. The result is called *orthometric height*. See Figure 12.1.

Indeed, elevation is still defined in this way. However, most measurements are made with GPS. The pre-satellite approach to measuring elevation is called *leveling*.

Pre-satellite measuring:

1. Start at sea level and call this zero elevation. (If there were no winds, currents, and tides, then the ocean surface would be an equipotential surface and all shorelines would be at exactly the same potential.)
2. Sight a line inland perpendicular to a plumb line. Note that this plumb line will be perpendicular to the equipotential surface and thus is not pointed toward the geocenter.
3. Measure the height difference and then move the setup inland and repeat the measurements until you reach the next shoreline. If all measurements are correct, you will be back to zero elevation assuming the ocean surface is an equipotential surface.

With artificial satellites, measuring geometric height is easier—especially if one is far from a coastline.

Measuring with artificial satellites:

1. Calculate the x, y, z position of each GPS satellite in the constellation using a global tracking network.
2. Measure the travel time to four or more satellites, three for position and one for clock error.
3. Establish your x, y, z position and convert this to *geometric height* above the spheroid, which we'll define below.
4. Go to a table of geoid height and subtract the local geoid height to get the orthometric height used by all surveyors and mappers.

Orthometric heights are useful because water flows downhill in this system, while it does not always flow downhill in the geometric height system. Of course the problem with orthometric heights is that they are very difficult to measure, or one must have a precise measurement of geoid height. Let *geodesists* worry about these issues.

12.1.2 Gravity

The second complication in the pre-satellite geodesy is the measurement of gravity. Interpretation of surface gravity measurement is either difficult or trivial, depending whether you are on land or at sea, respectively. Consider the land case illustrated in Figure 12.2.

Small variations in the acceleration of gravity ($<10^{-6}$ g) can be measured on the land surface. The major problem is that when the measurement is made in a

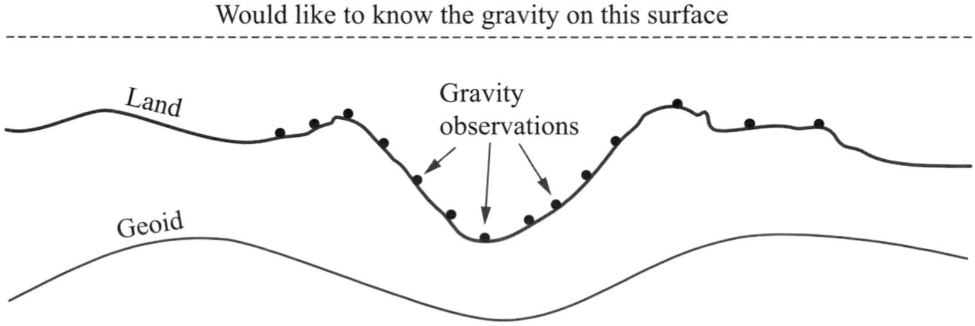

Figure 12.2

valley, there are masses above the observation plane. Thus, bringing the gravity measurement to a common level requires knowledge of the mass distribution above the observation point. This requires knowledge of both the geometric topography and the 3-D density. We could assume a constant density and use leveling to get the orthometric height, but we need to convert to geometric height to do the gravity correction. To calculate the geometric height, we need to know the geoid. However, the geoid height measurement comes from the gravity measurement, so there is no exact solution. Of course, one can make some approximations to get around this dilemma, but it is still a problem, and this is the fundamental reason why many geodesy books are so complicated. Of course, if one could make measurements of both the gravity and topography on a plane (or sphere) above all of the topography, our troubles would be over.

Ocean gravity measurements are much less of a problem, because the ocean surface is nearly equal to the geoid—so we can simply define the ocean gravity measurement as *free-air gravity*. We'll get back to all of this again later when we discuss flat-Earth approximations for gravity analysis.

12.2 Global Gravity

This section on global gravity is largely based on four books (Turcotte and Schubert (2014, Chapter 5, 5.1–5.5), Stacey (1977, Chapters 3–4), Jackson (1998, Chapter 3), Fowler (1990, Chapter 5)). The gravity field of the Earth can be decomposed as follows:

- the main field due to the total mass of the Earth;
- the second harmonic due to the flattening of the Earth by rotation; and
- anomalies which can be expanded in spherical harmonics or Fourier series.

The combined main field and the second harmonic make up the reference Earth model (i.e., spheroid, the reference potential, and the reference gravity). Deviations from this reference model are called elevation, geoid height, deflections of the vertical, and gravity anomalies.

12.2.1 Spherical Earth Model

The spherical Earth model is a good point to define some geodetic terms. There are both fundamental constants and derived quantities. See Table 12.1.

We should say a little more about units. Deviations in acceleration from the reference model, described next, are measured in units of milligal (1 mGal = 10^{-3} cm s^{-2} = 10^{-5} m s^{-2} = 10 gravity units (gu)). As noted above, the vertical

Table 12.1.

Parameter	Description	Formula	Value/Unit
R_e	mean radius of Earth	—	6371000 m
M_e	mass of Earth	—	5.98×10^{24} kg
G	gravitational constant	—	6.67×10^{-11} m^3 kg^{-1} s^{-2}
ρ	mean density of Earth	$M_e(4/3\pi R_e^3)^{-1}$	5520 kg m^{-3}
U	mean potential energy needed to take a unit mass from the surface of the Earth and place it at infinite distance	$-GM_e R_e^{-1}$	-6.26×10^7 m^2 s^{-2}
g	mean acceleration on the surface of the Earth	$-\delta U/\delta r = -GM_e R_e^{-2}$	-9.82 m s^{-2}
$\delta g/\delta r$	gravity gradient or free-air correction	$\delta g/\delta r = -2GM_e R_e^{-3}$ $= -2g/R_e$	3.086×10^{-6} s^{-2}

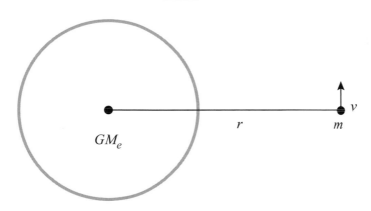

Figure 12.3

gravity gradient is also called the *free-air correction*, since it is the first term in the Taylor series expansion for gravity about the radius of the Earth.

$$g(r) = g(R_e) + \frac{\partial g}{\partial r}(r - R_e) + \cdots \tag{12.1}$$

Example 12.1 How does one measure the mass of the Earth? The best method is to time the orbital period of an artificial satellite. Indeed, measurements of all long-wavelength gravitational deviations from the reference model are best done by satellites. See Figure 12.3.

$$\frac{GM_e m}{r^2} = \frac{mv^2}{r}$$

$$\frac{GM_e}{r^2} = r\omega_s^2 \quad \Rightarrow \quad GM_e = r^3\omega_s^2$$

(12.2)

The mass is in orbit about the center of the Earth, so the outward centrifugal force is balanced by the inward gravitational force; this is Kepler's Third Law. If we measure the radius of the satellite orbit r and the orbital frequency ω_s, we can estimate GM_e. For example, the satellite Geosat has a orbital radius of 7168 km and a period of 6037.55 sec, so GM_e is 3.988708×10^{14} m^3 s^{-2}. Note that the product GM_e is tightly constrained by the observations, but that the accuracy of the mass of the Earth M_e is related to the accuracy of the measurement of G.

12.2.2 Ellipsoidal Earth Model

The centrifugal effect of the Earth's rotation causes an equatorial bulge that is the principal departure of the Earth from a spherical shape. If the Earth behaved like a fluid and there were no convective fluid motions, then it would be in hydrostatic equilibrium, and the Earth would assume the shape of an ellipsoid of revolution also called the *spheroid*. See Figure 12.4 and Table 12.2.

The formula for an ellipse in Cartesian coordinates is

$$\frac{x^2}{a^2} + \frac{y^2}{a^2} + \frac{z^2}{c^2} = 1$$

(12.3)

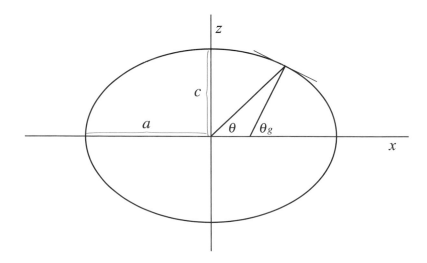

Figure 12.4

Table 12.2.

Parameter	Description	Formula	Value/Unit (WGS84)
GM_e			$3.986004418 \times 10^{14}$ m^3 s^{-2}
a	equatorial radius	—	6378137 m
c	polar radius	—	6356752.3 m
ω	rotation rate	—	7.292115×10^{-5} rad s^{-1}
f	flattening	$f = (a-c)/a$	1/298.257223563
θ_g	geographic latitude	—	—
θ	geocentric latitude	—	—

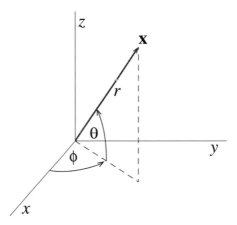

Figure 12.5

where the x-axis is in the equatorial plane at zero longitude (Greenwich), the y-axis is in the equatorial plane and at $90°\,E$ longitude, and the z-axis points along the spin-axis. The formula relating $x, y,$ and z to geocentric latitude and longitude is (see Figure 12.5)

$$x = r \cos\theta \, \cos\phi$$

$$y = r \cos\theta \, \sin\phi \qquad (12.4)$$

$$z = r \sin\theta.$$

Now we can rewrite the formula for the ellipse in polar coordinates and solve for the radius of the ellipse as a function of geocentric latitude.

$$r = \left(\frac{\cos^2\theta}{a^2} + \frac{\sin^2\theta}{c^2} \right)^{-1/2} \cong a \left(1 - f \sin^2\theta \right) \qquad (12.5)$$

Before satellites were available for geodetic work, one would establish geographic latitude by measuring the angle between a local plumb line and an external reference point, such as the star Polaris. Since the local plumb line is perpendicular to the spheroid (i.e., local flattened surface of the Earth), it points to one of the foci of the ellipse. The conversion between geocentric and geographic latitude is straightforward and its derivation is left as an exercise. The formulas are

$$\tan\theta = \frac{c^2}{a^2}\tan\theta_g \quad \text{or} \quad \tan\theta = (1-f)^2\tan\theta_g. \tag{12.6}$$

Example 12.2 What is the geocentric latitude at a geographic latitude of $45°$? The answer is $\theta = 44.8°$, which amounts to a 22 km difference in location!

12.2.3 Flattening of the Earth by Rotation

Suppose the Earth is a rotating, self-gravitating ball of fluid in hydrostatic equilibrium. Then density will increase with increasing depth and surfaces of constant pressure, and density will coincide. The surface of the Earth will be one of these equipotential surfaces; it has a potential U_o

$$U_o = V(r,\theta) - \tfrac{1}{2}\omega^2 r^2 \cos^2\theta \tag{12.7}$$

where the second term on the right side of equation (12.7) is the change potential due to the rotation of the Earth at a frequency ω. The potential due to an ellipsoidal Earth in a non-rotating frame can be expressed as

$$V = -\frac{GM_e}{r}\left[1 - J_1\frac{a}{r}P_1(\theta) - J_2\left(\frac{a}{r}\right)^2 P_2(\theta) - \cdots\right]. \tag{12.8}$$

The center of the coordinate system is selected to coincide with the center of mass, so by definition, J_1 is zero. For this model, we keep only J_2 (dynamic form factor or "jay two" $= 1.08 \times 10^{-3}$), so the final reference model is

$$V = -\frac{GM_e}{r} + \frac{GM_e J_2 a^2}{2r^3}\left(3\sin^2\theta - 1\right). \tag{12.9}$$

This parameter J_2 is related to the principal moments of inertia of the Earth by MacCullagh's formula. For a complete derivation see (Stacey, 1977, Chapter 3). Let C and A be the moments of inertia about the spin axis and equatorial axis, respectively. For example,

$$C = \int_V (x^2 + y^2)\,dm. \tag{12.10}$$

After a lot of algebra one can derive a relationship between J_2 and the moments of inertia.

$$J_2 = \frac{C - A}{Ma^2} \tag{12.11}$$

In addition, if we know J_2, we can approximately determine the flattening. This is done by inserting equation (12.9) into equation (12.7) and noting that the value of U_o is the same at the equator and the pole. Solving for the polar and equatorial radii that meet this constraint, one finds an approximate relationship between J_2 and the flattening.

$$f = \frac{a - c}{a} \cong \frac{3}{2} J_2 + \frac{1}{2} \frac{a^3 \omega^2}{GM} \tag{12.12}$$

Thus, if we could somehow measure J_2, we would know quite a bit about our planet.

12.2.4 Measurement of J_2

Just as in the case of measuring the total mass of the Earth, the best way to measure J_2 is to monitor the orbit of an artificial satellite. In this case, we measure the precessional period of the inclined orbit plane. To the second degree, external potential is

$$V = -\frac{GM_e}{r} + \frac{GM_e J_2 a^2}{2r^3} \left(3 \sin^2 \theta - 1 \right). \tag{12.13}$$

The force acting on the satellite is $-\nabla V$.

$$\mathbf{g} = -\frac{\partial V}{\partial r} \hat{r} - \frac{1}{r} \frac{\partial V}{\partial \theta} \hat{\theta} - \frac{1}{r \cos \theta} \frac{\partial V}{\partial \phi} \hat{\phi} \tag{12.14}$$

If we were out in space, the best way to measure J_2 would be to measure the $\hat{\theta}$-component of the gravity force.

$$g_\theta = \frac{1}{r} \frac{\partial V}{\partial \theta} = -\frac{3GM_e a^2 J_2}{r^4} \sin \theta \cos \theta \tag{12.15}$$

This component of force will apply a torque to the orbital angular momentum and it should be averaged over the orbit. Consider Figure 12.6.

For a prograde orbit, the precession ω_p is retrograde: the opposite to the Earth's spin direction. The complete derivation is found in Stacey (1977, page 76). The result is:

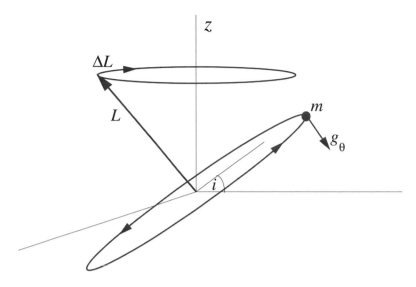

Figure 12.6 Satellite of mass m orbiting the Earth at an inclination i. The orbit has an angular momentum vector L perpendicular to the orbital plane. The θ component of the gravity vector applies a torque to the orbit, causing a retrograde precession of the angular momentum vector and thus a precession of the orbital plane.

$$\frac{\omega_p}{\omega_s} = \frac{-3a^2}{2r^2} J_2 \cos i \qquad (12.16)$$

where i is the inclination of the satellite orbit with respect to the equatorial plane, ω_s is the orbit frequency of the satellite, and ω_p is the precession frequency of the orbit plane in inertial space.

> **Example 12.3** LAGEOS: As an example, the LAGEOS satellite orbits the Earth every 13673.4 seconds, at an average radius of 12,265 km and an inclination of 109.8°. Given the parameters in Table 12.3 and that $J_2 = 1.08 \times 10^{-3}$, the predicted precession rate is 0.337°/day. This can be compared with the observed rate of 0.343°/day. Figure 12.7 is an illustration of the LAGEOS satellite.

12.2.5 Hydrostatic Flattening

Given the radial density structure, the Earth rotation rate and the assumption of hydrostatic equilibrium, one can calculate the theoretical flattening of the Earth (see Garland (1977, Appendix 2)). This is called the hydrostatic flattening $f_h = 1/299.5$. From Table 12.3, we have the observed flattening $f = 1/298.257$, so the actual Earth is flatter than the theoretical Earth. There are two reasons for this. First, the Earth is still recovering from the last ice age when the poles were loaded by heavy ice sheets. When the ice melted, polar dimples remained and the

Table 12.3. *LAGEOS Orbital Parameters.*

Description	Value
Semimajor axis	12,265 km
Eccentricity	0.004
Inclination	109.8°
Perigee height	5,858 km
Apogee height	5,958 km
Perigee rate	−0.215°/d
Node rate	+0.343°/d
Semimajor axis decay rate	−1.1 mm/d
Orbital acceleration	3×10^{-12} m s^{-2}

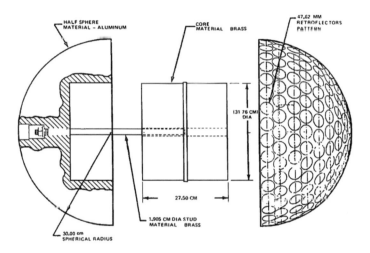

Figure 12.7 Structural detail of the LAGEOS satellite (Cohen and Smith, 1985).

glacial rebound of the viscous mantle is still incomplete. Second, the mantle is not in hydrostatic equilibrium, because of mantle convection. Finally, it should be noted that J_2 is changing with time due to the continual post-glacial rebound. This is called *jay two dot* $\dot{J}_2$ and it can be observed in satellite orbits as a time variation in the precession rate.

12.3 Exercises

Exercise 12.1 Derive equation (12.6) using the ellipsoidal Earth model.

Exercise 12.2 Derive equation (12.12) by following the instructions in the paragraph preceding this equation. Assume $\frac{a}{c} \cong 1$.

Exercise 12.3 Derive equation (12.16). You may need to refer to the book by Stacey (1977, Chapter 4, page 76).

Exercise 12.4 A sun synchronous orbit has a prograde precession rate of once per year. This type of orbit is used by optical remote sensing satellites to have the same sun illumumination for every repeat cycle. Use equation (12.16) to calculate the inclination of the orbit plane needed to match this rate for a satellite in a circular orbit at 800 km.

Exercise 12.5 Assume the density of the Earth is uniform and the earth is a perfect sphere.

(a) Develop a formula for the gravitational acceleration as a function of radius inside the Earth and check the dimensions.
(b) Develop a formula for the pressure at the center of this Earth and check the dimensions.

13

Reference Earth Model: WGS84

This short chapter provides a review of the four parameters that define the reference geometry and gravity model for the Earth called World Geodetic System 1984 (WGS84). Based on this reference model, anomalies are defined. These include geoid height, free-air gravity anomaly, and deflections of the vertical.

13.1 Some Definitions

Radius of spheroid

$$r(\theta) = \left(\frac{\cos^2 \theta}{a^2} + \frac{\sin^2 \theta}{c^2} \right)^{-1/2} \cong a(1 - f \sin^2 \theta) \tag{13.1}$$

Conversion between geocentric θ and geographic θ_g latitude

$$\tan \theta = \frac{c^2}{a^2} \tan \theta_g \quad \text{or} \quad \tan \theta = (1 - f)^2 \tan \theta_g \tag{13.2}$$

Gravitational potential in frame rotating with the Earth

$$U_o = -\frac{GM_e}{r} + \frac{GM_e J_2 a^2}{2r^3}(3 \sin^2 \theta - 1) - \tfrac{1}{2}\omega^2 r^2 \cos^2\theta \tag{13.3}$$

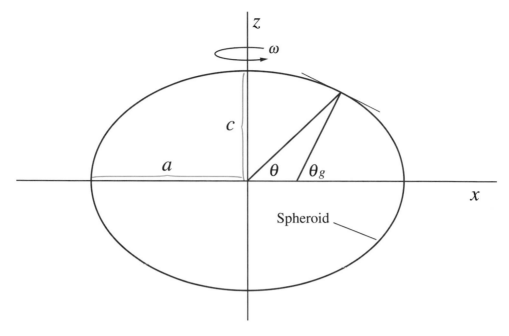

Figure 13.1

Calculation of the second degree harmonic J_2 from WGS84 parameters

$$J_2 \cong \frac{2}{3}f - \frac{a^3\omega^2}{3GM_e} \tag{13.4}$$

Calculation of J_2 from the polar C and equatorial A moments of inertia

$$J_2 = \frac{C - A}{M_e a^2} \tag{13.5}$$

Kepler's third law relating orbit frequency, ω_s, and radius r to M_e

$$\omega_s^2 r^3 = GM_e \tag{13.6}$$

Measurement of J_2 from orbit frequency ω_s radius r, inclination i, and precession rate ω_p

$$\frac{\omega_p}{\omega_s} = \frac{-3a^2}{2r^2} J_2 \cos i \tag{13.7}$$

Parameter	Description	Formula	Value/Unit
GM_e	(WGS84)	—	$3.986004418 \times 10^{14}$ m^3 s^{-2}
M_e	mass of Earth	—	5.98×10^{24} kg
G	gravitational constant	—	6.67×10^{-11} m^3 kg^{-1} s^{-2}
a	equatorial radius (WGS84)	—	6378137 m
c	polar radius (derived)	—	6356752.3 m
ω	rotation rate (WGS84)	—	7.292115×10^{-5} rad s^{-1}
f	flattening (WGS84)	$f = (a - c)/a$	1/298.257223563
J_2	dynamic form factor (derived)	—	1.081874×10^{-3}
θ_g	geographic latitude	—	—
θ	geocentric latitude	—	—

Hydrostatic flattening is less than observed flattening

$$f_H = \frac{1}{299.5} < f = \frac{1}{298.257} \tag{13.8}$$

13.2 Disturbing Potential and Geoid Height

To a first approximation, the reference potential U_o is constant over the surface of the Earth. Now, we are concerned with deviations from this reference potential. This is called the *disturbing potential* Φ; over the oceans the anomalous potential results in a deviation in the surface away from the spheroid

$$U = U_o + \Phi \tag{13.9}$$

where the reference potential U_o is given in equation (13.3). The *geoid* is the equipotential surface of the Earth that coincides with the sea surface when it is undisturbed by winds, tides, or currents. The *geoid height* N is the height of the geoid above the spheroid and it is expressed in meters. Consider the following mass anomaly in the Earth and its effect on the ocean surface.

Because of the excess mass, the potential on the spheroid is higher than the reference level $U = U_o + \Phi$. Thus, the ocean surface must move farther from

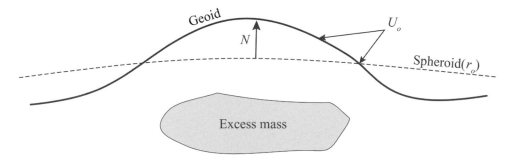

Figure 13.2

the center of the Earth to remain at the reference level U_o. To determine how far it moves, expand the potential in a Taylor series about the radius of the spheroid at r_o.

$$U_o(r) = U(r_o) + \frac{\partial U}{\partial r}(r - r_o) + \cdots \tag{13.10}$$

Notice that $g = -\delta U/\delta r$, so we arrive at

$$U(r) - U_o \cong g(r - r_o)$$

$$\Phi = gN. \tag{13.11}$$

This is *Brun's Formula*, which relates the disturbing potential to the geoid height N.

13.3 Reference Gravity and Gravity Anomaly

The *reference gravity* is the value of total (scalar) acceleration one would measure on the spheroid assuming no mass anomalies inside of the Earth.

$$\mathbf{g} = -\nabla U_o = -\frac{\partial U_o}{\partial r}\hat{r} - \frac{1}{r}\frac{\partial U_o}{\partial \theta}\hat{\theta} - \frac{1}{r\cos\theta}\frac{\partial U_o}{\partial \phi}\hat{\phi} \tag{13.12}$$

The total acceleration on the spheroid is

$$g = -\left[\left(\frac{\partial U_o}{\partial r}\right)^2 + \left(\frac{1}{r}\frac{\partial U_o}{\partial \theta}\right)^2\right]^{1/2}. \tag{13.13}$$

The second term on the right side of equation (13.13) is negligible, because the normal to the ellipsoid departs from the radial direction by a small amount, and the square of this value is usually unimportant. The result is

$$g(r,\theta) = -\frac{GM_e}{r^2}\left[1 - \frac{3\,J_2\,a^2}{2r^2}(3\sin^2\theta - 1)\right] + \omega^2\,r\cos^2\theta. \tag{13.14}$$

To calculate the value of gravity anomaly on the spheroid, we substitute

$$r(\theta) = a(1 - f\sin^2\theta) \tag{13.15}$$

After substitution, expand the gravity in a binomial series and keep terms of order f, but not f^2, and we arrive at the reference gravity on the spheroid.

$$g(\theta) = g_e \left[1 + \left(\tfrac{5}{2}m - f\right)\sin^2\theta\right]$$
$$m = \frac{\omega^2 a^2 c}{GM_e} \tag{13.16}$$

The parameter g_e is the value of gravity on the equator and m is approximately equal to the ratio of centrifugal force at the equator to the gravitational acceleration at the equator. In practice, geodesists get together at meetings of the International Union of Geodesy and Geophysics (IUGG) and agree on such things as the parameters of WGS84. In addition, they define something called the *international gravity formula*. This is also called the *Somigliana formula* in units of m s^{-2}.

$$g_o(\theta) = 9.780327 \left(1 + 5.3024 \times 10^{-3}\sin^2\theta - 5.8 \times 10^{-6}\sin^2 2\theta\right) \tag{13.17}$$

13.4 Free-Air Gravity Anomaly

The free-air gravity anomaly is the negative radial derivative of the disturbing potential, but it is also evaluated on the geoid. The formula is

$$\Delta g = -\frac{\partial \Phi}{\partial r} - \frac{2g_o(\theta)}{r(\theta)}N. \tag{13.18}$$

13.5 Summary of Anomalies

Disturbing potential Φ

$$\underset{\substack{\text{total}\\\text{potential}}}{U} = \underset{\substack{\text{reference}\\\text{potential}}}{U_o} + \underset{\substack{\text{disturbing}\\\text{potential}}}{\Phi} \tag{13.19}$$

Geoid height N

$$N = \frac{\Phi}{g_o(\theta)} \tag{13.20}$$

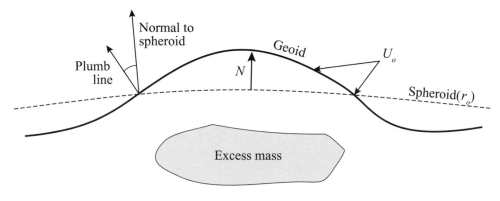

Figure 13.3

Free-air gravity anomaly

$$\Delta g = -\frac{\partial \Phi}{\partial r} - \frac{2g_o\,(\theta)}{r(\theta)}N \qquad (13.21)$$

Deflection of the vertical The final type of anomaly, not yet discussed, is the *deflection of the vertical*. This is the angle between the normal to the geoid (i.e., plumb line) and the normal to the spheroid. There are two components: north ξ and east η.

$$\xi = -\frac{1}{a}\frac{\partial N}{\partial \theta}$$

$$\eta = -\frac{1}{a\cos\theta}\frac{\partial N}{\partial \phi} \qquad (13.22)$$

14

Laplace's Equation in Spherical Coordinates

14.1 Introduction

As discussed in Chapter 12, the gravity field of the Earth can be decomposed into a reference gravity model (e.g., WGS84), and anomalies which can be expanded in spherical harmonics and/or Fourier series. The spherical harmonic decomposition should be used for longer wavelength anomalies (i.e., $\lambda > 1000$ km). However, for shorter wavelength anomalies (e.g., $\lambda < 1000$ km), the Fourier series representation is more practical and computationally efficient.

We begin by introducing spherical harmonics and their properties. We explain how the spherical harmonic decomposition of a function on a sphere is analogous to the Fourier series decomposition of a 2-D function in Cartesian coordinates. We then use this spherical harmonic formulation to solve Laplace's equation. Finally, we describe how the Earth's gravity field is represented as spherical harmonic coefficients. This chapter follows from Jackson (1998, Chapter 3). However, when we apply this mathematical development to the Earth, we replace the colatitude measured from the z-axis—commonly used by mathematicians and physicists— with the geocentric latitude measured from the equator—commonly used by Earth scientists.

14.2 Spherical Harmonics

The coordinate system used for this presentation of spherical harmonics follows from the definition used by mathematicians and physicists and is shown in Figure 14.1. Longitude is measured from the x-axis and colatitude θ is measured

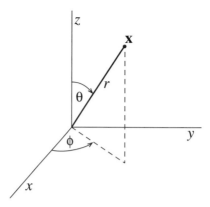

Figure 14.1

from the z-axis. Any function $f(\theta,\phi)$ can be expanded in terms of spherical harmonic coefficients as follows

$$f(\theta,\phi) = \sum_{l=0}^{\infty} \sum_{m=-l}^{m=l} F_l^m Y_l^m (\theta,\phi) \qquad (14.1)$$

where $Y_l^m (\theta,\phi)$ are the spherical harmonic functions, where l is the spherical harmonic degree and m is the spherical harmonic order. The spherical harmonic coefficients F_l^m are computed by integrating the function over the sphere as follows

$$F_l^m = \int_{0}^{2\pi} \int_{0}^{\pi} f(\theta,\phi)\overline{Y_l^m}(\theta,\phi) \sin\theta \; d\theta \; d\phi \qquad (14.2)$$

where the overbar signifies complex conjugate. The fully normalized spherical harmonic functions are

$$Y_l^m (\theta,\phi) = \left[\frac{(2l+1)(l-m)!}{4\pi(l+m)!} \right]^{1/2} P_l^m (\cos\theta) \, e^{im\phi}$$

$$Y_l^{-m} (\theta,\phi) = (-1)^m \, \overline{Y_l^m}(\theta,\phi) \qquad (14.3)$$

where $P_l^m (\cos\theta)$ is the associated Legendre function. Note that $e^{im\phi}$ represents the Fourier series that form a complete basis set of orthonormal functions of order m on the interval $0 \leqslant \phi \leqslant 2\pi$. Also, the functions $P_l^m (\cos\theta)$ form a complete basis set in the degree l for each m on the interval $0 \leqslant \theta \leqslant \pi$. Therefore, the $Y_l^m (\theta,\phi)$ functions form a complete orthonormal basis on the unit sphere. The orthonormal condition is

$$\int_{0}^{2\pi} \int_{0}^{\pi} Y_{l'}^{m'} (\theta,\phi) \, \overline{Y_l^m}(\theta,\phi) \sin\theta \; d\theta \; d\phi = \delta_{l'l}\,\delta_{m'm}. \qquad (14.4)$$

The completeness relation is

$$\sum_{l=0}^{\infty} \sum_{m=-l}^{m=l} \overline{Y_l^m}\left(\theta',\phi'\right) Y_l^m\left(\theta,\phi\right) = \delta\left(\phi - \phi'\right)\delta\left(\cos\theta - \cos\theta'\right). \qquad (14.5)$$

Some examples of spherical harmonic functions are

$$Y_0^0 = \frac{1}{\sqrt{4\pi}}$$

$$Y_1^{-1} = \sqrt{\frac{3}{8\pi}}\, \sin\theta e^{-i\phi}$$

$$Y_1^0 = \sqrt{\frac{3}{4\pi}}\, \cos\theta$$

$$Y_1^1 = -\sqrt{\frac{3}{8\pi}}\, \sin\theta e^{i\phi}$$

$$Y_2^{-2} = \sqrt{\frac{15}{32\pi}}\, \sin^2\theta e^{-i2\phi} \qquad (14.6)$$

$$Y_2^{-1} = \sqrt{\frac{15}{8\pi}}\, \sin\theta\cos\theta e^{-i\phi}$$

$$Y_2^0 = \sqrt{\frac{5}{16\pi}}\left(3\cos^2\theta - 1\right)$$

$$Y_2^1 = -\sqrt{\frac{15}{8\pi}}\, \sin\theta\cos\theta e^{i\phi}$$

$$Y_2^2 = \sqrt{\frac{15}{32\pi}}\, \sin^2\theta e^{i2\phi}.$$

The spherical harmonic order l is similar to the wavenumber in the Fourier series. For a sphere of radius a the characteristic wavelength of the spherical harmonic function is

$$\lambda = \frac{2\pi a}{(l+1)}. \qquad (14.7)$$

An examination of these spherical harmonic functions shows that there are always l nodes around the sphere. There are m nodes in longitude, so there are $l - m$ nodes in latitude. As discussed above, spherical harmonic decomposition of a function on a sphere is analogous to Fourier decomposition, as shown in the table below.

Cartesian Coordinates	Spherical Coordinates
$F(\mathbf{k}) = \int\limits_{-\infty}^{\infty} \int\limits_{-\infty}^{\infty} f(\mathbf{x}) e^{-i2\pi(\mathbf{k}\bullet\mathbf{x})} \, d^2\mathbf{x}$	$F_l^m = \int_0^{2\pi} \int_0^{\pi} f(\theta,\phi) \, \overline{Y_l^m}(\theta,\phi) \, \sin\theta \, d\theta \, d\phi$
$f(\mathbf{x}) = \int\limits_{-\infty}^{\infty} \int\limits_{-\infty}^{\infty} F(\mathbf{k}) e^{i2\pi(\mathbf{k}\bullet\mathbf{x})} \, d^2\mathbf{k}$	$f(\theta,\phi) = \sum_{l=0}^{\infty} \sum_{m=-l}^{m=l} F_l^m \, Y_l^m(\theta,\phi)$
$\mathrm{Re}\left[f(\mathbf{x})\right] \Rightarrow F\left(-k_x, k_y\right) = \overline{F}\left(k_x, k_y\right)$	$\mathrm{Re}\left[f(\theta,\phi)\right] \Rightarrow F_l^{-m} = (-1)^m \, \overline{F_l^m}$

The last row of the table shows the Hermitian property of the Fourier transform in the case where the function is real valued in the space domain. A similar property holds for spherical harmonic coefficients. In both cases, this property enables one to skip the computation of half of the coefficients—which saves both computer time and computer memory.

14.3 Laplace's Equation

Now we use this spherical harmonic decomposition to solve Laplace's equation in spherical coordinates. Similarly, in the following chapter, we will use the Fourier transform to solve Laplace's equation in Cartesian coordinates. Laplace's equation is

$$\frac{1}{r}\frac{\partial^2}{\partial r^2}(r\Phi) + \frac{1}{r^2\sin\theta}\frac{\partial}{\partial\theta}\left(\sin\theta\frac{\partial\Phi}{\partial\theta}\right) + \frac{1}{r^2\sin^2\theta}\frac{\partial^2\Phi}{\partial\phi^2} = 0 \qquad (14.8)$$

where $\Phi(r,\theta,\phi)$ is the potential, θ is colatitude, and ϕ is longitude. The boundary conditions are

$$\lim_{r\to\infty} \Phi(r,\theta,\phi) = 0$$
$$\qquad\qquad\qquad\qquad\qquad (14.9)$$
$$\Phi(1,\theta,\phi) = \Phi_o(\theta,\phi).$$

Now we expand the potential function in spherical harmonics

$$\Phi(r,\theta,\phi) = \sum_{l=0}^{\infty} \sum_{m=-l}^{m=l} \Phi_l^m(r) \, Y_l^m(\theta,\phi). \qquad (14.10)$$

With some work, it is possible to show that Laplace's equation reduces to

$$r\frac{\partial^2}{\partial r^2}(r\Phi_l^m) - l(l+1)\,\Phi_l^m = 0. \qquad (14.11)$$

The general solution to this ordinary differential equation is

$$\Phi_l^m(r) = A_l^m r^{-(l+1)} + B_l^m r^l. \tag{14.12}$$

Note that only the first term in this solution satisfies the boundary condition as r goes to infinity, so the solution is

$$\Phi(r,\theta,\phi) = \sum_{l=0}^{\infty} \sum_{m=-l}^{m=l} A_l^m \, r^{-(l+1)} \, Y_l^m(\theta,\phi). \tag{14.13}$$

The solution that satisfies the surface boundary condition is

$$\Phi(r,\theta,\phi) = \sum_{l=0}^{\infty} \sum_{m=-l}^{m=l} \Phi_{ol}^m \, r^{-(l+1)} \, Y_l^m(\theta,\phi) \tag{14.14}$$

where

$$\Phi_{ol}^m = \int_0^{2\pi} \int_0^{\pi} \Phi_o(\theta,\phi) \, \overline{Y_l^m}(\theta,\phi) \, \sin\theta \, d\theta \, d\phi. \tag{14.15}$$

This solution is used to calculate the potential anywhere exterior to the sphere using the following approach.

1. First expand the surface potential in spherical harmonics (equation (14.15)).
2. Then multiply each coefficient by the factor $r^{-(l+1)}$.
3. Finally, sum this product over all l and m (equation (14.13)).

The factor $r^{-(l+1)}$ is called the *upward continuation*. It reduces the amplitude of the potential—especially at large l, which corresponds to short wavelength. For the Cartesian coordinate system, the analogous upward continuation factor is $e^{-2\pi |k| z}$. In Figure 14.2, we compare the two upward continuation formulas to establish the wavelength where the Cartesian formulation can replace the spherical harmonic formulation.

14.4 Earth's Gravity Field

The spherical harmonic formulation commonly used for the Earth's gravity field uses geocentric latitude rather than colatitude. With this change, the fully normalized spherical harmonic function becomes

$$Y_l^m(\theta,\phi) = \left[\frac{(2l+1)(l-m)!}{4\pi(l+m)!}\right]^{1/2} P_l^m(\sin\theta) \, e^{im\phi} \tag{14.16}$$

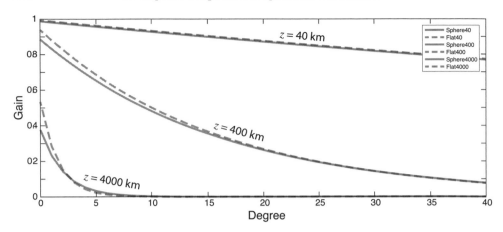

Figure 14.2 Amplitude gain versus spherical harmonic degree for spherical upward continuation (solid lines) compared with flat Earth upward continuation (dashed lines), for three altitudes: 40 km, 400 km, and 4000 km. For these altitudes, the two formulas show some disagreement for degrees less than 20 (1900 km wavelength), but good agreement for larger degrees. At degree 40, they are indistinguishable and the flat-Earth approximation is accurate enough for most applications.

where θ now refers to geocentric latitude, and there is a $\sin\theta$ in the associated Legendre function instead of a $\cos\theta$. As discussed in Chapter 12, the $l = 1$ spherical harmonic coefficient is zero, so the disturbing potential becomes

$$\Phi(r,\theta,\phi) = \frac{-GM_e}{r} \sum_{l=2}^{\infty} \sum_{m=0}^{m=l} A_l^m \left(\frac{a}{r}\right)^l Y_l^m(\theta,\phi) \qquad (14.17)$$

where GM_e and a are two components of the WGS84 reference model provided in Chapter 13. Note that the C_2^0 term is sometimes equivalent to the J_2 term used in the reference model 13.3. This formulation has complex valued coefficients A_l^m, but it is customary to use real coefficients C_l^m and S_l^m applied to the sine and cosine components of the complex exponential. Moreover, the standard gravity approach has a different normalization and uses the fact that the gravity field is real valued to eliminate the need for negative order coefficients. With these changes, the disturbing potential is

$$\Phi(r,\theta,\phi) = \frac{-GM_e}{r} \sum_{l=2}^{\infty} \sum_{m=0}^{m=l} \left(\frac{a}{r}\right)^l \tilde{P}_l^m(\sin\theta) \left[C_l^m \cos(m\phi) + S_l^m \sin(m\phi)\right]$$

$$(14.18)$$

where this associated Legendre function has a slightly different normalization:

$$\tilde{P}_l^m(\sin\theta) = \left[\frac{(2-\delta_{0,m})\,(2l+1)\,(l-m)!}{(l+m)!}\right]^{1/2} P_l^m(\sin\theta), \tag{14.19}$$

One can relate the real valued spherical harmonic coefficients used for Earth gravity analysis to the complex coefficients used by mathematical physicists as

$$A_l^m = \begin{cases} \sqrt{2\pi}\left(C_l^m - iS_l^m\right)(-1)^m & m > 0 \\ \sqrt{4\pi}\,C_l^m & m = 0 \\ \sqrt{2\pi}\left(C_l^m + iS_l^m\right) & m < 0. \end{cases} \tag{14.20}$$

In practice, geodesists have a standard way of distributing the spherical harmonic coefficients. Below are the tide-free coefficients up to degree 6, for the Earth Gravity model 2008 (EGM2008) gravity model, from Pavlis et al. (2012).

degree	order	Clm	Slm	sigClm	sigSlm
2	0	-0.484165143790815D-03	0.000000000000000D+00	0.7481239490D-11	0.0000000000D+00
2	1	-0.206615509074176D-09	0.138441389137979D-08	0.7063781502D-11	0.7348347201D-11
2	2	0.243938357328313D-05	-0.140027370385934D-05	0.7230231722D-11	0.7425816951D-11
3	0	0.957161207093473D-06	0.000000000000000D+00	0.5731430751D-11	0.0000000000D+00
3	1	0.203046201047864D-05	0.248200415856872D-06	0.5726633183D-11	0.5976692146D-11
3	2	0.904787894809528D-06	-0.619005475177618D-06	0.6374776928D-11	0.6401837794D-11
3	3	0.721321757121568D-06	0.141434926192941D-05	0.6029131793D-11	0.6028311182D-11
4	0	0.539965866638991D-06	0.000000000000000D+00	0.4431111968D-11	0.0000000000D+00
4	1	-0.536157389388867D-06	-0.473567346518086D-06	0.4568074333D-11	0.4684043490D-11
4	2	0.350501623962649D-06	0.662480026275829D-06	0.5307840320D-11	0.5186098530D-11
4	3	0.990856766672321D-06	-0.200956723567452D-06	0.5631952953D-11	0.5620296098D-11
1	4	-0.188519633023033D-06	0.308803882149194D-06	0.5372877167D-11	0.5383247677D-11
5	0	0.686702913736681D-07	0.000000000000000D+00	0.2910198425D-11	0.0000000000D+00
5	1	-0.629211923042529D-07	-0.943698073395769D-07	0.2989077566D-11	0.3143313186D-11
5	2	0.652078043176164D-06	-0.323353192540522D-06	0.3822796143D-11	0.3642768431D-11
5	3	-0.451847152328843D-06	-0.214955408306046D-06	0.4725934077D-11	0.4688985442D-11
5	4	-0.295328761175629D-06	0.498070550102351D-07	0.5332198489D-11	0.5302621028D-11
5	5	0.174811795496002D-06	-0.669379935180165D-06	0.4980396595D-11	0.4981027282D-11
6	0	-0.149953927978527D-06	0.000000000000000D+00	0.2035490195D-11	0.0000000000D+00
6	1	-0.759210081892527D-07	0.265122593213647D-07	0.2085980159D-11	0.2193954647D-11
6	2	0.486488924604690D-07	-0.373789324523752D-06	0.2603949443D-11	0.2466506184D-11
6	3	0.572451611175653D-07	0.895201130010730D-08	0.3380286162D-11	0.3347204566D-11
6	4	-0.860237937191611D-07	-0.471425573429095D-06	0.4535102219D-11	0.4489428324D-11
6	5	-0.267166423703038D-06	-0.536493151500206D-06	0.5097794605D-11	0.5101153019D-11
6	6	0.947068749756882D-08	-0.237382353351005D-06	0.4731651005D-11	0.4728357086D-11

In summary, one can calculate the gravity field at a radius greater than the equatorial radius of the Earth. If the calculation is performed in an Earth-fixed coordinate system rotating with the Earth, then the gravity computed using the WGS84 reference model in equation (13.3) is added to the disturbing potential computed using spherical harmonic coefficients; see equation (14.18).

$$U(r,\theta,\phi) = \frac{-GM_e}{r}\left\{1 + \sum_{l=2}^{\infty}\sum_{m=0}^{m=l}\left(\frac{a}{r}\right)^l \tilde{P}_l^m(\sin\theta)\left[C_l^m\cos(m\phi) + S_l^m\sin(m\phi)\right]\right\}$$

$$-\frac{1}{2}\omega^2 r^2\cos^2\theta$$

(14.21)

There is an excellent web service (`icgem.gfz-potsdam.de`) where one can select their favorite gravity model as spherical harmonic coefficients and then compute a wide variety of gravity products (e.g., geoid height, free-air anomaly, gravity gradient, ...) on a grid defined by the user.

14.5 Exercises

Exercise 14.1 Show that equation (14.12) is a solution to the differential equation (14.11).

Exercise 14.2 Write a MATLAB program to generate Figure 14.2. Also plot the ratio of the two upward continuation functions to degree 80.

15

Laplace's Equation in Cartesian Coordinates and Satellite Altimetry

15.1 Solution to Laplace's Equation

Variations in the gravitational potential and in the gravitational force are caused by local variations in the mass distribution in the Earth. As described in Chapter 12 we decompose the gravity field of the Earth into three fields:

- the main field due to the total mass of the Earth;
- the second harmonic due to the flattening of the Earth by rotation; and
- anomalies which can be expanded in spherical harmonics or Fourier series.

Here we are interested in anomalies due to local structure. Consider a patch on the Earth having a width and length less than about 1000 km, or 1/40 of the circumference of the Earth. Within that patch, we are interested in features as small as perhaps 1 km wavelength. Using a spherical harmonic representation would require 40,000 squared coefficients! To avoid this enormous computation and still achieve accurate results, we will treat the Earth as being locally flat. Here is a remove/restore approach that has worked well in our analysis of gravity and topography:

1. Acquire a spherical harmonic model of the gravitational potential of the Earth and generate models of the relevant quantities (e.g., geoid height, gravity anomaly, deflection of the vertical, ...) out to harmonic 80. You may want to taper the harmonics between, say, 60 and 120 to avoid Gibbs phenomenon; this depends on the application.
2. *Remove* that model from the local geoid, gravity, ... An alternate method is to remove a trend from the data and then apply some type of window prior to performing the Fourier analysis. I do not recommend this practice because the trend being removed will contain a broad spectrum; it is dependent on the size of the area, and it cannot be restored accurately.
3. Project the residual data onto a Mercator grid so that the cells are approximately square, and then use the central latitude of the grid to establish the dimensions of the grid for Fourier analysis.

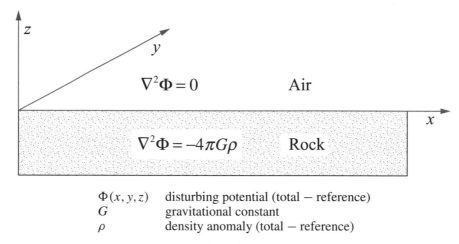

$\Phi(x, y, z)$ disturbing potential (total − reference)
G gravitational constant
ρ density anomaly (total − reference)

Figure 15.1

4. Perform the desired calculation (e.g., upward continuation, gravity/topography transfer function, ...).
5. *Restore* the appropriate spherical harmonic quantity using the exact model that was removed originally.

Consider the disturbing potential

$$U \;\; = \;\; U_o \;\; + \;\; \Phi \tag{15.1}$$

total
potential   reference
potential   disturbing
potential

where, in this case, the reference potential comprises the reference Earth model plus the reference spherical harmonic model described in Step 1 above. The disturbing potential satisfies Laplace's equation for an altitude, z, above the highest mountain in the area, while it satisfies Poisson's equation below this level, as shown in Figure 15.1.

Laplace's equation is a second order partial differential equation in three dimensions.

$$\frac{\partial^2 \Phi}{\partial x^2} + \frac{\partial^2 \Phi}{\partial y^2} + \frac{\partial^2 \Phi}{\partial z^2} = 0, \quad z > 0 \tag{15.2}$$

Six conditions are needed to develop a unique solution. Far from the region, the disturbing potential must go to zero; this accounts for five of the boundary conditions.

$$\lim_{|x|\to\infty} \Phi = 0, \qquad \lim_{|y|\to\infty} \Phi = 0, \qquad \lim_{z\to\infty} \Phi = 0 \qquad (15.3)$$

At the surface of the Earth (or at some elevation), one must either prescribe the potential or the vertical derivative of the potential.

$$\Phi(x,y,0) = \Phi_o(x,y) \quad \textit{(Dirichlet)}$$

$$\frac{\partial \Phi}{\partial z} = -\Delta g(x,y) \quad \textit{(Neumann)} \qquad (15.4)$$

To solve this differential equation, we'll use the 2-D Fourier transform again, where the forward and inverse transform are

$$F(\mathbf{k}) = \int\limits_{-\infty}^{\infty}\int\limits_{-\infty}^{\infty} f(\mathbf{x}) e^{-i2\pi(\mathbf{k}\cdot\mathbf{x})} d^2\mathbf{x}$$

$$f(\mathbf{x}) = \int\limits_{-\infty}^{\infty}\int\limits_{-\infty}^{\infty} F(\mathbf{k}) e^{i2\pi(\mathbf{k}\cdot\mathbf{x})} d^2\mathbf{k} \qquad (15.5)$$

where $\mathbf{x} = (x,y)$ is the position vector, $\mathbf{k} = (1/\lambda_x, 1/\lambda_y)$ is the wavenumber vector, and $\mathbf{k}\cdot\mathbf{x} = k_x x + k_y y$. Fourier transformation reduces Laplace's equation and the surface boundary to

$$-4\pi^2\left(k_x^2 + k_y^2\right)\Phi(\mathbf{k},z) + \frac{\partial^2\Phi}{\partial z^2} = 0 \qquad (15.6)$$

$$\lim_{z\to\infty} \Phi(\mathbf{k},z) = 0, \qquad \Phi(\mathbf{k},0) = \Phi_o.$$

The general solution is

$$\Phi(\mathbf{k},z) = A(\mathbf{k})e^{2\pi|\mathbf{k}|z} + B(\mathbf{k})e^{-2\pi|\mathbf{k}|z}. \qquad (15.7)$$

To satisfy the boundary condition as $z \to \infty$, the $A(k)$ term must be zero. To satisfy the boundary condition on the $z = 0$ plane, $B(k)$ must be $\Phi(k,0)$. The final result is

$$\underset{\substack{\text{potential at}\\\text{altitude}}}{\Phi(\mathbf{k},z)} = \underset{\substack{\text{potential at}\\ z=0}}{\Phi_o(\mathbf{k},0)} \times \underset{\substack{\text{upward}\\\text{continuation}}}{e^{-2\pi|\mathbf{k}|z}}. \qquad (15.8)$$

The upward continuation physics, as shown in Figure 15.2, is the same for the potential and all of its derivatives. For example, if one measured gravity anomaly at the surface of the Earth $\Delta g(x,0)$, then to compute the gravity at an altitude of z, one takes the Fourier transform of the surface gravity, multiplies it by the upward continuation kernel, and inverse transforms the result. This exponential decay of

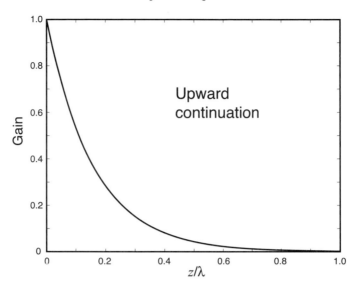

Figure 15.2 Gain of upward continuation kernel as a function of the altitude of the observation z, divided by the wavelength of the anomaly λ.

the signal with altitude is a fundamental barrier to recovery of small-scale gravity anomalies from a measurement made at altitude. Here are two important examples:

1. *Marine gravity*   Consider making marine gravity measurement on a ship which is 4 km above the topography of the ocean floor. (Most of the short-wavelength gravity anomalies are generated by the mass variations associated with the topography of the seafloor.) At a wavelength of 8 km, the ocean surface anomaly will be attenuated by 0.043 from the amplitude of the seafloor anomaly.

2. *Satellite gravity*   The typical altitude of an artificial satellite used to sense variations in the gravity field is 400 km, so an anomaly having a 100 km wavelength will be attenuated by a factor of 10^{-11}! This is why radar altimetry (below), which measures the geoid height directly on the ocean surface topography, is so valuable.

15.2 Derivatives of the Gravitational Potential

This solution to Laplace's equation can be used to construct all of the common derivatives of the potential. Suppose one has a complete survey over a patch on the surface of the Earth so that a Fourier method can be used to convert between the different representations of the gravity field. This is particularly true for computing gravity anomaly from geoid height or deflection of the vertical. The general relation between the potential in the space domain (at any altitude) and the Fourier transform of the surface potential is

$$\Phi(\mathbf{x}, z) = \int\limits_{-\infty}^{\infty}\int\limits_{-\infty}^{\infty} \Phi(\mathbf{k}, 0) e^{-2\pi|\mathbf{k}|z} e^{i2\pi(\mathbf{k}\cdot\mathbf{x})} d^2\mathbf{k}. \tag{15.9}$$

Table 15.1 uses equation (15.9) and the definitions of the derivatives of the potential to construct the variety of anomalies. Before examining these relationships, however, let's review some of the definitions in relation to what can be measured.

Gravitational Potential

N: **Geoid height** Since the ocean surface is an equipotential surface, variations in gravitational potential will produce variations in the sea surface height. This can be measured by a radar altimeter.

First Derivative of Potential

Δg: **Gravity anomaly** This is the derivative of the potential with respect to z. It can be measured by an accelerometer, such as a gravity meter.

η, ξ: **Deflection of the vertical** These are the derivatives of the potential with respect to x and y, respectively. They can be measured by recording the tiny angle between a plumb bob and the vector pointing to the center of the Earth. Over the ocean, this is most easily measured by taking the along-track derivative of radar altimeter profiles.

Second derivative of potential

Gravity gradient This is a symmetric tensor of second partial derivatives of the gravitational potential.

$$\begin{bmatrix} \dfrac{\partial^2 \Phi}{\partial x^2} & \dfrac{\partial^2 \Phi}{\partial x \partial y} & \dfrac{\partial^2 \Phi}{\partial x \partial z} \\[2ex] & \dfrac{\partial^2 \Phi}{\partial y^2} & \dfrac{\partial^2 \Phi}{\partial y \partial z} \\[2ex] & & \dfrac{\partial^2 \Phi}{\partial z^2} \end{bmatrix}$$

Table 15.1. *Relationships between the various representations of the gravity field in free space.*

	Space domain	Wavenumber domain
Geoid height from the potential, Bruns formula	$N(\mathbf{x}) \cong \frac{1}{g}\Phi(\mathbf{x},0)$	$N(\mathbf{k}) \cong \frac{1}{g}\Phi(\mathbf{k},0)$ (15.10)
Gravity anomaly from the potential	$\Delta g(\mathbf{x},z) \cong -\dfrac{\partial \Phi}{\partial z}(\mathbf{x},z)$	$\Delta g(\mathbf{k},z) \cong 2\pi\,|\mathbf{k}|\,e^{-2\pi|\mathbf{k}|z}\,\Phi(\mathbf{k},0)$ (15.11)
Deflection of the vertical from the potential (east slope and north slope)	$\eta(\mathbf{x}) = -\dfrac{\partial N}{\partial x} \cong -\dfrac{1}{g}\dfrac{\partial \Phi}{\partial x}$ $\xi(\mathbf{x}) = -\dfrac{\partial N}{\partial y} \cong -\dfrac{1}{g}\dfrac{\partial \Phi}{\partial y}$	$\eta(\mathbf{k}) \cong -\dfrac{i2\pi k_x}{g}\,\Phi(\mathbf{k},0)$ $\xi(\mathbf{k}) \cong -\dfrac{i2\pi k_y}{g}\,\Phi(\mathbf{k},0)$ (15.12)
Gravity anomaly from deflection of the vertical (Haxby et al., 1983)		$\Delta g(\mathbf{k}) = \dfrac{ig}{|\mathbf{k}|}\left[k_x\eta(\mathbf{k}) + k_y\xi(\mathbf{k})\right]$ (15.13)
Vertical gravity gradient from the curvature of the ocean surface	$\dfrac{\partial \Delta g}{\partial z} = g\left(\dfrac{\partial^2 N}{\partial x^2} + \dfrac{\partial^2 N}{\partial y^2}\right)$ (15.14)	

A direct way of making this measurement is to construct a set of accelerometers, each spaced at a distance of Δ in the $x, y,$ and z directions to measure each component of gravity gradient. Note that when the gravity gradient measurements are made in free space, the trace of this tensor must be zero by Laplace's equation. Below we'll use Laplace's equation to develop an alternate method of measuring $\frac{\partial^2 \Phi}{\partial z^2}$ over the ocean using a radar altimeter.

As an exercise, use Laplace's equation and the various definitions to develop gravity anomaly from vertical deflection (equation (15.13)) and vertical gravity gradient from ocean surface curvature (equation (15.14)).

Here is a practical example: Suppose one has measurements of geoid height $N(\mathbf{x})$ over a large area on the surface of the ocean and wishes to calculate the gravity anomaly, $\Delta g(\mathbf{x}, z)$ at altitude. The prescription is:

1. Remove an appropriate spherical harmonic model from the geoid.
2. Take the 2-D Fourier transform of $N(\mathbf{x})$.
3. Multiply by $g 2\pi \, |\mathbf{k}| \, e^{-2\pi|\mathbf{k}|}$.
4. Take the inverse 2-D Fourier transform.
5. Restore the matching gravity anomaly calculated from the spherical harmonic model at altitude.

15.3 Geoid Height, Gravity Anomaly, and Vertical Gravity Gradient from Satellite Altimeter Profiles

As described above, geoid height $N(\mathbf{x})$ and other measurable quantities such as gravity anomaly $g(\mathbf{x})$ are related to the anomalous gravitational potential $\Phi(\mathbf{x}, z)$ through Laplace's equation. It is instructive to go through an example of how measurements of ocean surface topography from satellite radar altimetry can be used to construct geoid height, deflection of the vertical, gravity anomaly, and vertical gravity gradient.

The surface of the ocean is displaced both above and below the reference ellipsoidal shape of the Earth (Figure 15.3). These differences in height arise from variations in gravitational potential (i.e., geoid height) and oceanographic effects (tides, large-scale currents, el Niño, eddies, . . .). Fortunately, the oceanographic effects are small compared with the permanent gravitational effects, so a radar altimeter can be used to measure these bumps and dips. At wavelengths less than about 200 km, bumps and dips in the ocean surface topography reflect the topography of the ocean floor and can be used to estimate seafloor topography in areas of sparse ship coverage.

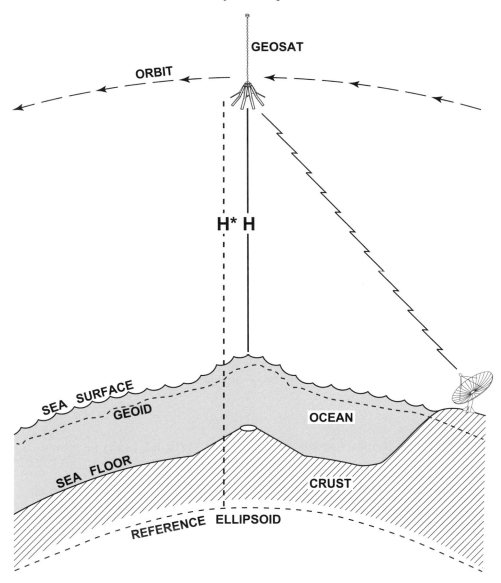

Figure 15.3 Schematic diagram of a radar altimeter orbiting the Earth at a typical altitude of 800 km.

Radar altimeters are used to measure the height of the ocean surface above the reference ellipsoid (i.e., in Figure 15.3 the satellite above the ellipsoid H^* minus the altitude above the ocean surface H). A GPS, or ground-based tracking system, is used to establish the position of the radar H^* (as a function of time) to an accuracy of better than 0.1 m. The radar emits 1000 pulses per second, using a carrier wavelength of about 2 cm (Ku-band). These spherical wave fronts reflect

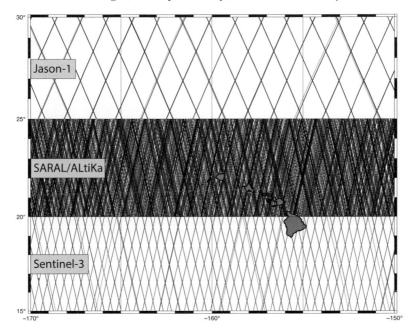

Figure 15.4 Ground tracks for three radar altimeters that have provided gravity coverage of the ocean surface. This is a 2000 km by 1500 km area around Hawaii. For most of its lifetime the Jason-1 altimeter was in an orbit that repeated every 10 days. This is optimal for observing changes in sea surface height associated with oceanographic processes but inadequate for gravity field recovery. After the launch of Jason-2 (not shown) the Jason-1 track was shifted to bisect the coverage. The lower tracks are from the twin Sentinel-3 satellite altimeters which provide repeat coverage every 27 days. The center tracks are from the SARAL ALtiKa altimeter which operated in a 35-day repeat cycle and then was placed in a drifting orbit that achieves excellent coverage for gravity recovery. To date (2020) there have been nine altimeters with dense geodetic coverage including Geosat, ERS-1, Envisat, Jason-1/2 (extension of life), CryoSat-2, SARAL/ALtiKa and Sentinel-3, and HY-2.

from the closest ocean surface (nadir) and return to the satellite, where the two-way travel times is recorded to an accuracy of 3 nanoseconds (1 m range variations mostly due to ocean waves). Averaging thousands of pulses reduces the noise to about 30 mm. If one is interested in making an accurate geoid height map such as shown in Figure 15.5, then many sources of error must be considered and somehow removed. However, if the final product of interest is one of the derivatives of the potential, then it is best to take the along-track derivative of each profile to develop along-track sea surface slope. In this case, the point-to-point precision of the measurements is the limiting factor.

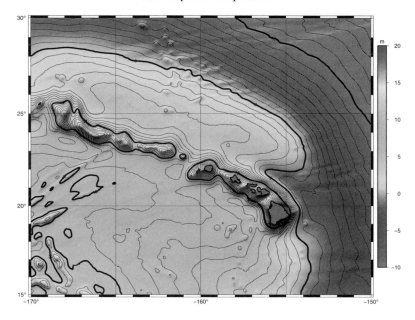

Figure 15.5 Geoid height above the WGS84 ellipsoid in meters (1 m contour interval) from Earth Gravity model 2008 (EGM2008) (Pavlis et al., 2012). The geoid height is dominated by long wavelengths, so it is difficult to observe the small-scale features caused by ocean-floor topography. These can be enhanced by computing either the horizontal derivative (ocean surface slope) or the vertical derivative (gravity anomaly).

To avoid a crossover adjustment of the data, ascending and descending satellite altimeter profiles are first differentiated in the along-track direction, resulting in geoid slopes or along-track vertical deflections. These along-track slopes are then combined to produce east η and north ξ components of vertical deflection. Finally, the east and north vertical deflections are used to compute both gravity anomaly and vertical gravity gradient. The details for converting along-track slope into east and north components of deflection of the vertical are provided in Section 15.4 and also in Sandwell and Smith (1997). You probably don't need to know these details unless you plan to do research in marine gravity.

To compute the gravity anomaly (Figure 15.6) from a dense network of satellite altimeter profiles of geoid height (Figure 15.3), one constructs grids of east η and north ξ vertical deflection (Figures 15.7 and 15.8). The grids are then Fourier transformed and equation (15.13) is used to compute the gravity anomaly. At this point, one can add the long wavelength gravity field from the spherical harmonic

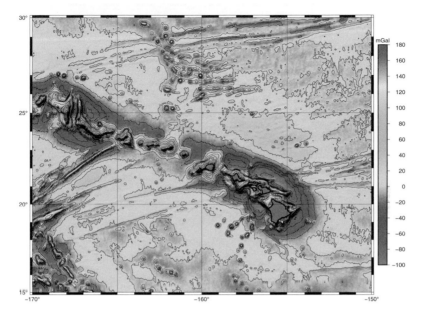

Figure 15.6 Gravity anomaly $\Delta g(x)$ derived from east and north components of sea surface slope using equation (15.13). (20 mGal contour interval.)

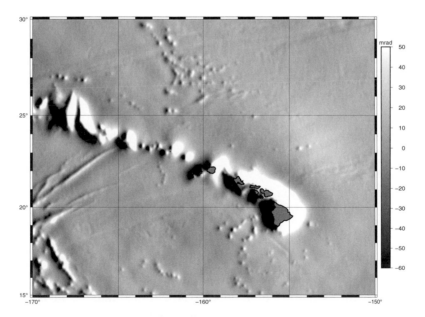

Figure 15.7 East component of sea surface slope $\eta(x)$ derived from satellite radar altimeter profiles. Note this component is rather noisy, because the altimeter tracks (Figure 15.4) run mainly in a N-S direction.

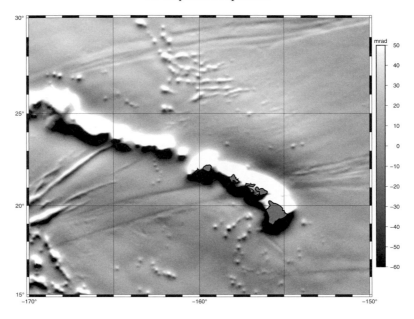

Figure 15.8 North component of sea surface slope $\xi(x)$ derived from satellite radar altimeter profiles. Note this component has lower noise, because the altimeter tracks (Figure 15.4) run mainly in a N-S direction.

model to the gridded gravity values in order to recover the total field; the resulting sum may be compared with gravity measurements made on board ships. A more complete description of gravity field recovery from satellite altimetry can be found in Hwang and Parsons (1996); Sandwell and Smith (1997); Rapp and Yi (1997).

There is an important issue for constructing the gravity anomaly from sea surface slope that is revealed by a simplified version of equation (15.13). Consider a 2-D sea surface slope and gravity anomaly which depends on x, but not y. The y-component of slope is zero, so conversion from sea surface slope to gravity anomaly is simply a Hilbert transform

$$\Delta g(k_x) = i g \, \text{sgn}(k_x) \eta \, (k_x) \, . \tag{15.15}$$

Now it is clear that one μrad of sea surface slope maps into 0.98 mGal of gravity anomaly and similarly one μrad of slope error will map into $\sim$1 mGal of gravity anomaly error. Thus, the accuracy of the gravity field recovery is controlled by the accuracy of the sea surface slope measurement.

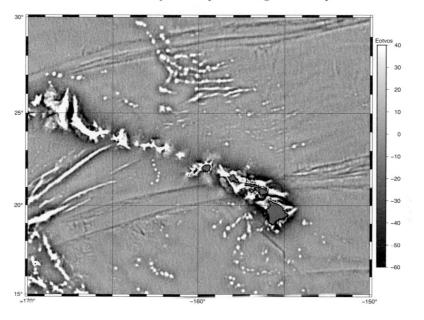

Figure 15.9 Vertical gravity gradient $\delta g(x)/\delta z$ derived from east and north components of sea surface slope using equation (15.14). Note this second derivative of the geoid amplifies the shortest wavelengths (compare with the original geoid (Figure 15.5)). Noise in the altimeter measurements has been amplified, resulting in an artificial texture.

15.4 Vertical Deflections from Along-Track Slopes

15.4.1 Crossover Method

Consider for the moment the intersection point of an ascending and a descending satellite altimeter profile. The derivative of the geoid height N with respect to time t along the ascending profile is

$$\dot{N}_a = \frac{\partial N_a}{\partial t} = \frac{\partial N}{\partial \theta}\dot{\theta}_a + \frac{\partial N}{\partial \phi}\dot{\phi}_a \qquad (15.16)$$

and along the descending profile is

$$\dot{N}_d = \frac{\partial N}{\partial \theta}\dot{\theta}_d + \frac{\partial N}{\partial \phi}\dot{\phi}_d \qquad (15.17)$$

where θ is geodetic latitude and ϕ is longitude. The functions θ and ϕ are the latitudinal and longitudinal components of the satellite ground track velocity. It is assumed that the satellite altimeter has a nearly circular orbit, so that its velocity depends mainly on latitude; at the crossover point, the following relationships are accurate to better than 0.1%.

$$\dot{\theta}_a = -\dot{\theta}_d \quad \dot{\phi}_a = \dot{\phi}_d \tag{15.18}$$

The geoid gradient (deflection of the vertical) is obtained by solving equation (15.16) and equation (15.17), using equation (15.18):

$$\frac{\partial N}{\partial \phi} = \frac{1}{2\dot{\phi}}(\dot{N}_a + \dot{N}_d) \tag{15.19}$$

$$\frac{\partial N}{\partial \theta} = \frac{1}{2|\dot{\theta}|}(\dot{N}_a - \dot{N}_d). \tag{15.20}$$

It is evident from this formulation that there are latitudes where either the east or north component of geoid slope may be poorly determined. For example, at $\pm 72°$ latitude, the Seasat and Geosat altimeters reach their turning points where the latitudinal velocity θ goes to zero and thus equation (15.20) becomes singular. In the absence of noise, this is not a problem, because the ascending and descending profiles are nearly parallel so their difference goes to zero at the same rate that the latitudinal velocity goes to zero. Of course in practice, altimeter profiles contain noise such that the north component of geoid slope will have a signal-to-noise ratio that decreases near $\pm 72°$ latitude. Similarly, for an altimeter in a near polar orbit, the ascending and descending profiles are nearly anti-parallel at the low latitudes; the east component of geoid slope is poorly determined and the north component is well determined. The optimal situation occurs when the tracks are nearly perpendicular so that the east and north components of geoid slope have the same signal-to-noise ratio.

When two or more satellites with different orbital inclinations are available, the situation is slightly more complex, but also more stable. Consider the intersection of four passes as shown in Figure 15.10. The along-track derivative of each pass can be computed from the geoid gradient at the crossover point

$$\begin{bmatrix} \dot{N}_1 \\ \dot{N}_2 \\ \dot{N}_3 \\ \dot{N}_4 \end{bmatrix} = \begin{bmatrix} \dot{\theta}_1 & \dot{\phi}_1 \\ \dot{\theta}_2 & \dot{\phi}_2 \\ \dot{\theta}_3 & \dot{\phi}_3 \\ \dot{\theta}_4 & \dot{\phi}_4 \end{bmatrix} \begin{bmatrix} \frac{\partial N}{\partial \theta} \\ \frac{\partial N}{\partial \phi} \end{bmatrix} \tag{15.21}$$

or in matrix notation

$$\dot{N} = \Theta \, \Delta N. \tag{15.22}$$

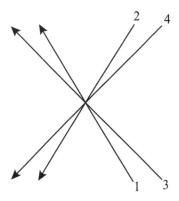

Figure 15.10

Since this is an overdetermined system, the four along-track slope measurements cannot be matched exactly unless the measurements are error-free. In addition, an *u priori* estimate of the error in the along-track slope σ_i, measurements can be used to weight each equation in equations (15.21) (i.e., divide each of the four equations by σ_i). The least squares solution to equation (15.22) is

$$\Delta N = (\Theta^t \, \Theta)^{-1} \, \Theta^t \, \dot{N} \qquad (15.23)$$

where t and -1 are the transpose and inverse operations, respectively. In this case, a 2-by-4 system must be solved at each crossover point, although the method is easily extended to three or more satellites. Later we will assume that every grid cell corresponds to a crossover point of all the satellites considered, so this small system must be solved many times.

In addition to the estimates of geoid gradient, the covariances of these estimates are also obtained.

$$\begin{bmatrix} \sigma^2_{\theta\theta} & \sigma^2_{\theta\phi} \\ \sigma^2_{\phi\theta} & \sigma^2_{\phi\phi} \end{bmatrix} = (\Theta^t \, \Theta)^{-1} \qquad (15.24)$$

Since Geosat and ERS-1 are high inclination satellites, the estimated uncertainty of the east component is about 3 times greater than the estimated uncertainty of the north component at the equator. At higher latitudes of 60°–70°, where the tracks are nearly perpendicular, the north and east components are equally well determined. At 72° north, where the Geosat tracks run in a westerly direction, the uncertainty of the east component is low and the higher inclination ERS-1 tracks prevent the estimate of the north component from becoming singular at 72°.

Finally, the east η and north ξ components of vertical deflection are related to the two geoid slopes by

$$\eta = -\frac{1}{a \cos \theta} \frac{\partial N}{\partial \phi} \tag{15.25}$$

$$\xi = -\frac{1}{a} \frac{\partial N}{\partial \theta} \tag{15.26}$$

where a is the mean radius of the Earth.

15.4.2 Biharmonic Splines

The approach provided in the previous section only works at crossover points and the actual spatial distribution of along-track slopes from several satellite altimeters is non-uniform. A more flexible approach is to find the smoothest surface that is compatible with all the slope data. Consider N estimates of slope $s(\mathbf{x}_i)$ with direction $\mathbf{n}_i$ each having uncertainty σ_i where $\mathbf{x} = (x, y)$. We wish to find the smoothest surface $w(\mathbf{x})$ that is consistent with this set of data such that $s_i = (\nabla w \cdot \mathbf{n})_i$. We develop a smooth model using a thin elastic plate that is subjected to vertical point loads. The loads are located at the locations of the data constraints (knots) and their amplitudes are adjusted to match the observed slopes (Sandwell, 1987). To suppress overshooting oscillations a tension can be applied to its perimeter. Wessel and Bercovici (1998) solved this problem by determining the Green's function $\phi(\mathbf{x})$ for the deflection of a thin elastic plate in tension. The differential equation is

$$\alpha^2 \nabla^4 \phi(\mathbf{x}) - \nabla^2 \phi(\mathbf{x}) = \delta(\mathbf{x}) \tag{15.27}$$

where α is a length scale factor that controls the importance of the tension. High α results in biharmonic spline interpolation which minimizes the strain energy in the plate but can produce undesirable oscillations between data points. These can be suppressed using a singular value decomposition method when solving the linear system of equations (below). Zero α corresponds to harmonic interpolation, which results in a surface that has sharp local perturbations at the locations of the data constraints. Thus, the tension factor controls the shape of the interpolating surface. Through experimentation we find good-looking results when the solution is about 0.33 of the way from the biharmonic to the harmonic end-member. The Green's function for this differential operator is (Wessel and Bercovici, 1998)

$$\phi(\mathbf{x}) = K_o \left(\frac{|\mathbf{x}|}{\alpha} \right) + \log \left(\frac{|\mathbf{x}|}{\alpha} \right) \tag{15.28}$$

where K_o is the modified Bessel function of the second kind and order zero. In the special case of zero tension, the Green's function is (Greenberg, 2015)

$$\phi(\mathbf{x}) = |\mathbf{x}|^2 \ln |\mathbf{x}|. \tag{15.29}$$

The smooth surface is a linear combination of these Green's functions each centered at the location of the data constraint.

$$w(\mathbf{x}) = \sum_{j=1}^{N} c_j \phi(\mathbf{x} - \mathbf{x_j}) \tag{15.30}$$

The coefficients c_j represent the strength of each point load applied to the thin elastic plate. They are found by solving the following linear system of equations.

$$s_i = (\nabla w \cdot \mathbf{n})_i = \sum_{j-1}^{N} c_j \nabla \phi(\mathbf{x}_i - \mathbf{x}_j) \cdot \mathbf{n}_i \tag{15.31}$$

If one is also interested in recovering the mean sea surface height ($\sim$geoid height) then additional height constraints should be added to the linear system of equations

$$w_i = \sum_{j=1}^{M} c_j \phi(\mathbf{x}_i - \mathbf{x}_j). \tag{15.32}$$

One issue that must be addressed is the possibility of having multiple constraints in exactly (or nearly) the same location. This causes the linear system to be exactly singular (or numerically unstable). Satellite altimeter data commonly have many crossing profiles so it is possible to have two or even ten slope constraints at nearly the same location. The solution to this problem is to reduce the number of Green's functions (knots) by making sure they are not more closely spaced than some prescribed distance. That minimum distance should be about a quarter of the shortest wavelength that one strives to resolve. When the number of knot locations is less than the number of data constraints, then the linear system is over-determined and the surface will not exactly match the slope constraints. Since we only wish to match the slopes to within the expected uncertainty of each data type, each equation (15.31) and (15.32), should be divided by the slope and height uncertainty to provide the optimal solution using a singular value decomposition algorithm. For recovery of the gravity field and its derivatives, we are not usually interested in the absolute height of the surface but just the local slope, so our final result is the gradient of the surface. For the zero tension, biharmonic case, the gradient is

$$\nabla \phi(\mathbf{x}) = \mathbf{x}(2 \ln |\mathbf{x}| + 1). \tag{15.33}$$

If we are interested in mean sea surface height ($\sim$geoid height) then equation (15.30) should be used. While this interpolation theory is elegant and very flexible, it is difficult to apply to the altimeter interpolation problem because today (2020) there are over 3 billion observations to grid. Consider gridding just 1000 slopes, the matrix of the linear system could have 10^6 elements if all the knot points were retained. In practice, we make the following compromises in order to grid this large and diverse set of data. (1) The data are residuals with respect to a model (e.g., EGM2008) so we can assemble and grid the data in overlapping small areas. (2) To avoid edge effects the sub areas have 100% overlap and only the inner interpolated cells are retained. (3) The along-track slope data from each of the ten possible slope directions (i.e. ascending and descending profiles from five satellite inclinations ERS/Envisat/ALtiKa, GEOSAT, TOPEX/Jason-1/2, CryoSat-2, and Sentinel-3) and associated uncertainties are binned onto the regularly spaced, square grid cells (1 minute or smaller), and only the median slope of each type is retained for fitting. The results of the computations are grids of residual east and north vertical deflection that are converted to gravity anomalies and vertical gravity gradient using equations (15.12) and (15.13), respectively. Note this gridding approach is available in Generic Mapping Tools (GMT) as the function *greenspline* and also in MATLAB as the function *griddata*.

15.5 Exercises

Exercise 15.1 Use Laplace's equation and the various definitions to develop gravity anomaly from vertical deflection (equation (15.13)) and vertical gravity gradient from ocean surface curvature (equation (15.14)).

Exercise 15.2 Show that equation (15.29) is the Green's function for the biharmonic equation by showing the following equation is true $\nabla^4 |\mathbf{x}|^2 \ln |\mathbf{x}| = 8\delta(\mathbf{x})$.

16

Poisson's Equation in Cartesian Coordinates

16.1 Solution to Poisson's Equation

As in Chapter 15 on Laplace's equation, we are interested in anomalies due to local structure and will use a flat-Earth approximation. However, unlike Chapter 15, the emphasis is on generating models of the disturbing potential and its derivatives from a 3-D model of the variations in density and topography of the Earth. In Chapter 17, we'll combine this approach to calculating gravity models with the models for isostasy and flexure, to develop a topography-to-gravity transfer function. Consider the disturbing potential

$$U = U_o + \Phi \qquad (16.1)$$

total   reference   disturbing
potential   potential   potential

where, in this case, the reference potential comprises the ellipsoidal reference Earth model plus the reference spherical harmonic model. The disturbing potential satisfies Laplace's equation for an altitude z above the highest mountain in the area, while it satisfies Poisson's equation below this level, as shown in Figure 16.1.

First, consider a density model consisting of an infinitesimally thin sheet at a depth z_o having a surface-density of $\sigma(x, y)$ (units of mass per unit area). Later, we'll construct a more complicated 3-D structure from a stack of many layers. Poisson's equation is an inhomogeneous second-order partial differential equation in three dimensions.

$$\frac{\partial^2 \Phi}{\partial x^2} + \frac{\partial^2 \Phi}{\partial y^2} + \frac{\partial^2 \Phi}{\partial z^2} = -4\pi \, G\sigma(\mathbf{x}) \, \delta(z - z_o) \qquad (16.2)$$

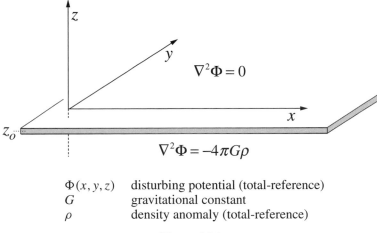

$\Phi(x, y, z)$ disturbing potential (total-reference)
G gravitational constant
ρ density anomaly (total-reference)

Figure 16.1

Six conditions are needed to develop a unique solution. Far from the region, the disturbing potential must go to zero; this accounts for the five boundary conditions.

$$\lim_{|x| \to \infty} \Phi = 0, \quad \lim_{|y| \to \infty} \Phi = 0, \quad \lim_{z \to \infty} \Phi = 0 \qquad (16.3)$$

The sixth condition is prescribed by the density model. Fourier transformation reduces Poisson's equation and the surface boundary to

$$-4\pi^2 \left(k_x^2 + k_y^2 \right) \Phi(\mathbf{k}, z) + \frac{\partial^2 \Phi}{\partial z^2} = -4\pi \, G\sigma(\mathbf{k}) \, \delta(z - z_o) \qquad (16.4)$$

$$\lim_{z \to \infty} \Phi(\mathbf{k}, z) = 0. \qquad (16.5)$$

Next, take the Fourier transform with respect to z

$$\pi \left(k_x^2 + k_y^2 + k_z^2 \right) \Phi(\mathbf{k}, k_z) = G\sigma(\mathbf{k}) e^{-i2\pi k_z z_o}. \qquad (16.6)$$

We have used the definition of the delta function

$$\int_{-\infty}^{\infty} \delta(z - z_o) e^{-i2\pi kz} \, \mathrm{d}z = e^{-i2\pi kz_o}.$$

Next, we solve the differential equation for Φ and take the inverse Fourier transform with respect to k_z

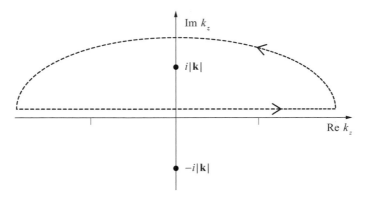

Figure 16.2

$$\Phi(\mathbf{k}, z) = \frac{G\sigma(\mathbf{k})}{\pi} \int\limits_{-\infty}^{\infty} \frac{e^{i2\pi k_z(z-z_o)}}{k_z^2 + \left(k_x^2 + k_y^2\right)} \, dk_z. \tag{16.7}$$

Use calculus of residues to do the integration. The denominator can be factored as

$$k_z^2 + \left(k_x^2 + k_y^2\right) = (k_z + i \, |\mathbf{k}|)(k_z - i \, |\mathbf{k}|) \tag{16.8}$$

where $|\mathbf{k}| = \left(k_x^2 + k_y^2\right)^{\frac{1}{2}}$. If $z > z_o$, then to satisfy the boundary condition as $z \to \infty$, one must integrate around the $i \, |\mathbf{k}|$-pole. See Figure 16.2.

The result is

$$\int\limits_{-\infty}^{\infty} \frac{e^{i2\pi k_z(z-z_o)}}{(k_z + i \, |\mathbf{k}|)(k_z - i \, |\mathbf{k}|)} \, dk_z = 2\pi i \frac{e^{-2\pi |\mathbf{k}|(z-z_o)}}{2i \, |\mathbf{k}|}. \tag{16.9}$$

The solution for the potential for $z > z_o$ is

$$\Phi(\mathbf{k}, z) = G\sigma(\mathbf{k}) \frac{e^{-2\pi |\mathbf{k}|(z-z_o)}}{|\mathbf{k}|}. \tag{16.10}$$

The gravity anomaly is

$$\Delta g(\mathbf{k}, z) = -\frac{\partial \Phi}{\partial z} = 2\pi G\sigma(\mathbf{k}) e^{-2\pi |\mathbf{k}|(z-z_o)}. \tag{16.11}$$

Ocean surface

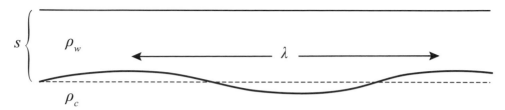

Figure 16.3

16.2 Gravity Due to Seafloor Topography: Approximate Formula

Consider topography on the ocean floor $t(\mathbf{x})$ where the maximum amplitude of the topography is much less than the mean ocean depth s, as shown in Figure 16.3.

Because the topography has low amplitude, we can replace the surface density in equation (16.11) with the topography times the density contrast across the seafloor.

$$\Delta g(\mathbf{k}) = 2\pi G \, (\rho_c - \rho_w) \, T(\mathbf{k}) e^{-2\pi |\mathbf{k}| s} \tag{16.12}$$

The result shows that, to a first approximation, the relationship between gravity and topography is linear and isotropic.

$$\frac{\Delta g}{T} = 2\pi G \, (\rho_c - \rho_w) \, e^{-2\pi |\mathbf{k}| s} \tag{16.13}$$

At long wavelengths, $|\mathbf{k}| \to 0$, so the exponential upward continuation term is 1 and the gravity/topography ratio is simply the Bouguer correction term.

$$\frac{\Delta g}{T} = 2\pi G \, (\rho_c - \rho_w) = 75 \text{ mGal/km} \tag{16.14}$$

Suppose the wavelength of the topography is equal to the ocean depth. In this case, the exponential, upward continuation reduces the gravity measured on the ocean surface by a factor of $e^{-2\pi} = 0.0017$. Because of this upward continuation, topography having wavelength less than the ocean depth is difficult to observe in the gravity field at the ocean surface.

16.3 Gravity Anomaly from a 3-D Density Model

Using this formulation, one can stack, or integrate, these surface density layers over a range of depths to construct the gravity field due to a full 3-D density model. See Figure 16.4.

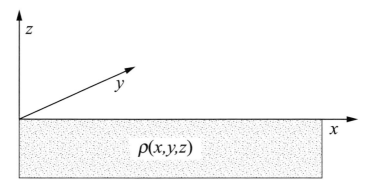

Figure 16.4

$$\Phi(\mathbf{k}, z) = G \int_{-\infty}^{o} \rho(\mathbf{k}, z_o) \frac{e^{-2\pi |\mathbf{k}|(z-z_o)}}{|\mathbf{k}|} \, dz_o \qquad (16.15)$$

The equivalent expression in the space domain is

$$\Phi(\mathbf{x}, z) =$$

$$G \int_{-\infty}^{\infty} \int_{-\infty}^{\infty} \int_{-\infty}^{o} \rho(x_o, y_o, z_o) \left[(x - x_o)^2 + (y - y_o)^2 + (z - z_o)^2 \right]^{-1/2} dz_o \, dy_o \, dx_o.$$

$$(16.16)$$

Indeed, this is just a statement of the convolution theorem where

$$\Im \left[\left(x^2 + y^2 + z^2 \right)^{-1/2} \right] = \frac{e^{-2\pi |\mathbf{k}|z}}{|\mathbf{k}|}. \qquad (16.17)$$

16.4 Computation of Geoid Height and Gravity Anomaly

Table 16.1 provides the two approaches for calculating geoid height and gravity anomaly from a 3-D density model. The Fourier approach involves 2-D Fourier transformation of each layer, summing (or integrating) the upward-continued contribution from each layer, and inverse Fourier transformation of the total. The space-domain approach involves a 3-D convolution of the density model with the $1/r$ (geoid) or z/r^3 (gravity) kernel. For a model with 1024 points in both horizontal directions, the Fourier approach will be about 50,000 times faster to compute than the space-domain convolution. Moreover, the Fourier approach will have higher numerical accuracy, because there are fewer additions and subtractions.

Table 16.1.

Space Domain	Wavenumber Domain
$N(\mathbf{x}) = \dfrac{G}{g} \displaystyle\int_{-\infty}^{\infty}\int_{-\infty}^{\infty}\int_{-\infty}^{o} \dfrac{\rho\,(x_o, y_o, z_o)}{\left[(x - x_o)^2 + (y - y_o)^2 + z_o^2\right]^{1/2}}\,dz_o\,dy_o\,dx_o$	$N(\mathbf{k}) = \dfrac{G}{g}\displaystyle\int_{-\infty}^{o}\rho(\mathbf{k}, z_o)\dfrac{e^{2\pi|\mathbf{k}|z_o}}{|\mathbf{k}|}\,dz_o$
$\Delta g(\mathbf{x}) = G \displaystyle\int_{-\infty}^{\infty}\int_{-\infty}^{\infty}\int_{-\infty}^{o} \dfrac{\rho\,(x_o, y_o, z_o)\,z_o}{\left[(x - x_o)^2 + (y - y_o)^2 + z_o^2\right]^{3/2}}\,dz_o\,dy_o\,dx_o$	$\Delta g(\mathbf{k}) = 2\pi G \displaystyle\int_{-\infty}^{o}\rho(\mathbf{k}, z_o)e^{2\pi|\mathbf{k}|z_o}\,dz_o$

16.5 Gravity Anomaly for a Slab: Bouguer Anomaly

The equation relating gravity to the 3-D density anomaly in the wavenumber
domain can be used to calculate the gravity anomaly due to a slab of thickness H
and a density of ρ_o. This is used for the Bouguer correction in land gravity surveys.
The 3-D density is

$$\rho(\mathbf{x}, z) = \begin{cases} \rho_o & -H < z < 0 \\ 0 & z < -H,\ z > 0. \end{cases} \tag{16.18}$$

The Fourier transform of this density is

$$\rho(\mathbf{k}, z) = \begin{cases} \delta\,(k_x)\,\delta\left(k_y\right)\rho_o & -H < z < 0 \\ 0 & z < -H,\ z > 0. \end{cases} \tag{16.19}$$

The gravity anomaly integral simplifies to

$$\Delta g(\mathbf{k}) = 2\pi\,G\rho_o\,\delta\,(k_x)\,\delta\left(k_y\right)\int_{-H}^{o} e^{2\pi|\mathbf{k}|z_o}\,dz_o \tag{16.20}$$

$$= 2\pi\,G\rho_o\delta\,(k_x)\,\delta\left(k_y\right)\frac{1}{2\pi\,|\mathbf{k}|}\left(1 - e^{-2\pi|\mathbf{k}|H}\right).$$

Since only the zero wavenumber component is extracted by the delta function, we
expand equation (16.20) in a Taylor series about $|\mathbf{k}|$ and take the limit as $|\mathbf{k}| \to 0$.

$$\lim_{|\mathbf{k}|\to 0}\frac{1}{2\pi\,|\mathbf{k}|}\left[1 - 1 + 2\pi\,|\mathbf{k}|\,H - \frac{(2\pi\,|\mathbf{k}|\,H)^2}{2!} + \cdots\right] = H \tag{16.21}$$

The result in the wavenumber domain is

$$\Delta g(\mathbf{k}) = 2\pi\,G\rho_o\,\delta\,(k_x)\,\delta\left(k_y\right)H. \tag{16.22}$$

The inverse Fourier transform provides the gravity field due to an infinite slab.

$$\Delta g(\mathbf{x}) = 2\pi \, G\rho_o H \qquad (16.23)$$

Over the ocean, one measures the total acceleration of gravity and subtracts the International Gravity Formula (IGF) to obtain the free-air gravity anomaly. Indeed, the free-air anomaly is defined on the geoid, which is closely approximated by the ocean surface. Therefore, no corrections are needed for marine gravity measurements.

In contrast, over the land, one measures total gravitational acceleration at some elevation h above the geoid; assume this elevation is known from leveling. To reduce these gravity measurements to the geoid, two corrections are commonly applied:

1. The free-air correction accounts for the decrease in gravity, because the observation point is farther from the center of the Earth.
2. The Bouguer correction uses the infinite-slab approximation to account for the gravitational attraction of the rock between the measurement point and the geoid. Note that unless the topography is very flat over a large area, this infinite-slab approximation may not be very accurate, and a more accurate terrain correction should be applied.

$$\Delta g_B \;=\; g_t \;-\; 2\pi G \rho_o h \;+\; \frac{2G M_e}{R_e^3} h \;-\; \gamma_o(\theta) \qquad (16.24)$$

| Bouguer gravity | measured gravity | slab correction -0.1118 mGal m^{-1} ($\rho_o = 2670$ kg m^{-3}) | free-air correction 0.3086 mGal m^{-1} | International Gravity Formula |

16.6 Gravity Anomaly from Topography: Parker's Exact Formula

In the previous development (Section 16.2), we collapsed the topography $t(\mathbf{x})$ into a thin sheet with varying surface density. The approximation is accurate when the amplitude of the topography is less than the upward continuation distance. For example, in the ocean where the mean seafloor depth is 4 km, the approximation works quite well for topography that extends 2 km above that depth. However, when the topography is rugged and approaches the observation plane (e.g., sea surface), a more accurate treatment is needed. Parker (1973) derived a more accurate formula that results in a Taylor series expansion in powers of topography.

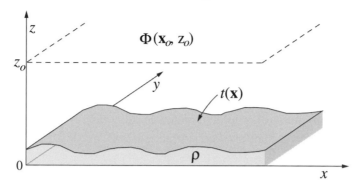

Figure 16.5

Consider the exact formula for the disturbing potential $\Phi(\mathbf{x_o}, z_o)$ due to a uniform density slab having a flat bottom and upper surface topography $t(\mathbf{x})$ (see Figure 16.5). This is given by the following 3-D convolution integral

$$\Phi(\mathbf{x_o}, z_o) = \rho G \int\limits_{-\infty}^{\infty} \int\limits_{-\infty}^{\infty} \int\limits_{0}^{t(\mathbf{x})} \left[(x-x_o)^2 + (y-y_o)^2 + (z-z_o)^2\right]^{-1/2} dz \, d^2\mathbf{x}.$$

(16.25)

Take the 2-D Fourier transform of the disturbing potential on the observation plane

$$\Im[\Phi] = \Phi(\mathbf{k}, z_o) = \rho G \int\limits_{-\infty}^{\infty} \int\limits_{-\infty}^{\infty} \int\limits_{0}^{t(\mathbf{x})} \int\limits_{-\infty}^{\infty} \int\limits_{-\infty}^{\infty} \left[(x-x_o)^2 + (y-y_o)^2 + (z-z_o)^2\right]^{-1/2}$$

$$\times e^{-i2\pi(\mathbf{k}\cdot\mathbf{x_o})} d^2\mathbf{x_o} \, dz \, d^2\mathbf{x}.$$

(16.26)

In Section 16.3, we showed that the 2-D Fourier transform of convolution the inverse distance Green's function is given by equation (16.17):

$$\Im\left[(x^2 + y^2 + z^2)^{-1/2}\right] = \frac{e^{-2\pi|\mathbf{k}|z}}{|\mathbf{k}|}.$$

(16.27)

Also, recall the shift property of the Fourier transform equation (2.23):

$$\Im\left[f(\mathbf{x}-\mathbf{x_o})\right] = e^{-i2\pi(\mathbf{k}\cdot\mathbf{x})} F(\mathbf{k}).$$

(16.28)

Using these tools, we can write the Fourier transform of the disturbing potential as

$$\Phi\left(\mathbf{k}, z_o\right) = \rho G \int_{-\infty}^{\infty} \int_{-\infty}^{\infty} \int_{0}^{t(\mathbf{x})} \frac{e^{-2\pi|k|(z_o-z)}}{|k|} e^{-i2\pi(\mathbf{k}\cdot\mathbf{x})} \, dz \, d^2\mathbf{x}. \tag{16.29}$$

The integral over z can be performed analytically.

$$e^{-2\pi|\mathbf{k}|z_o} \int_{0}^{t(\mathbf{x})} e^{2\pi|\mathbf{k}|z} dz = \frac{e^{-2\pi|\mathbf{k}|z_o}}{2\pi|\mathbf{k}|} \left[e^{2\pi|\mathbf{k}|t(\mathbf{x})} - 1 \right] \tag{16.30}$$

We can expand the term in brackets in a Taylor series about $|\mathbf{k}| = 0$.

$$\left[1 + 2\pi|\mathbf{k}|\, t\left(\mathbf{x}\right) + \frac{|2\pi\mathbf{k}|^2}{2!} t^2\left(\mathbf{x}\right) + \cdots - 1 \right] \tag{16.31}$$

Now we can rewrite the Fourier transform of the disturbing potential on the plane as

$$\Phi\left(\mathbf{k}, z_o\right) = 2\pi\rho G e^{-2\pi|k|z_o} \sum_{n=1}^{\infty} \frac{|2\pi\mathbf{k}|^{n-2}}{n!} \Im\left[t^n\left(\mathbf{x}\right)\right]. \tag{16.32}$$

Finally, note that the gravity anomaly is the negative vertical derivative of the potential $\Delta g = -\frac{\partial \Phi}{\partial z}$, so the result is

$$\Delta g\left(\mathbf{k}, z_o\right) = 2\pi\rho G e^{-2\pi|k|z_o} \sum_{n=1}^{\infty} \frac{|2\pi\mathbf{k}|^{n-1}}{n!} \Im\left[t^n\left(\mathbf{x}\right)\right]. \tag{16.33}$$

This exact formula for computing gravity anomaly from topography involves an infinite series of Fourier transforms of the topography raised to the power n. In the derivation of the approximate formula for gravity due to seafloor topography (Section 16.2), we compressed the topography, times the density contract across the seafloor, into surface density at an average seafloor depth. We see now that this is equivalent to keeping just the $n = 1$ term in equation (16.33) to arrive at

$$\Delta g\left(\mathbf{k}, z_o\right) = 2\pi\rho G e^{-2\pi|k|z_o} T\left(\mathbf{k}\right). \tag{16.34}$$

Parker (1973) proves that this series converges as long as the highest peak in the topography does not extend above the observation plane. Moreover, the convergence of the series is optimal when the $z = 0$ level is selected such that it is half way between the maximum and minimum topography.

One can use this more exact formula for calculating gravity due to flexurally compensated topography, as discussed in the next chapter.

16.7 Exercises

Exercise 16.1 Abyssal hills on the seafloor have a characteristic wavelength of 10 km and a peak-to-trough amplitude of 500 m.

(a) What is the amplitude of the gravity anomaly on the seafloor assuming the topography (density 2800 kg m^{-3}) can be compressed into a thin sheet?
(b) What is the amplitude of the gravity anomaly at the sea surface where the mean ocean depth is 3 km?
(c) Over a time period of 50 Ma, the abyssal hills will be carried by plate tectonics into the deep ocean where the depth is 5 km. What is the new amplitude of the gravity anomaly?
(d) In addition to a deeper ocean, the topography of the abyssal hills will be covered with sediment so the seafloor is now flat. What is the new value of the amplitude of the gravity? (Use a sediment density of 2300 kg m^{-3}.)

Exercise 16.2 Derive the Bouguer formula (equation 16.23) for the gravity due to a slab of uniform thickness H and uniform density ρ_o from the Parker expansion for gravity due to topography of uniform density (equation (16.33)). Replace $t(\mathbf{x})$ by H and continue the calculation.

17

Gravity/Topography Transfer Function and Isostatic Geoid Anomalies

17.1 Introduction

This chapter combines thin-elastic plate flexure theory with the solution to Poisson's equation, to develop a linear relationship between gravity and topography. This relationship can be used in a variety of ways:

1. If both the topography and gravity are measured over an area that is several times greater than the flexural wavelength, then the gravity/topography relationship (in the wavenumber domain) can be used to estimate the elastic thickness of the lithosphere and/or the crustal thickness. There are many good references on this topic, including: Dorman and Lewis (1972); McKenzie and Bowin (1976); Banks et al. (1977); Watts (1978); McNutt (1979).
2. At wavelengths greater than the flexural wavelength, where features are isostatically compensated, the geoid/topography ratio can be used to estimate the depth of compensation of crustal plateaus and the depth of compensation of hot-spot swells (Haxby and Turcotte, 1978).
3. If the gravity field is known over a large area, but there is rather sparse ship-track coverage, the topography/gravity transfer function can be used to interpolate the seafloor depth among the sparse ship soundings (Smith and Sandwell, 1994).

17.2 Flexure Theory

In Chapter 8, we developed an analytic solution for the response of a thin-elastic plate floating on a fluid mantle that is subjected to a line load. Here we follow the same approach, but solve the flexure equation for an arbitrary vertical load representing, for example, the loading of the lithosphere due to the weight of a volcano. This is shown in Figure 17.1.

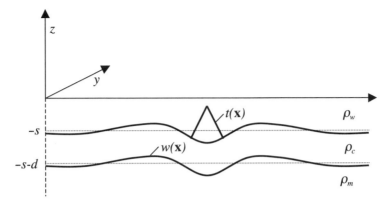

Figure 17.1

where s is the mean ocean depth ($\sim$4km) and d is the thickness of the crust ($\sim$6km). The topography of the Moho is equal to the deflection of the elastic plate $w(\mathbf{x})$. The topography of the seafloor, $t(\mathbf{x})$, has two components: the topographic load, $t_o(\mathbf{x})$, and the deflection of the elastic plate $w(\mathbf{x})$:

$$t(\mathbf{x}) = t_o(\mathbf{x}) + w(\mathbf{x}) \tag{17.1}$$

For this calculation, we make the following assumptions: the thickness of the elastic plate is less than the flexural wavelength; the deflection of the elastic plate is much less than the flexural wavelength; the flexural rigidity, D, is constant; and there is no end-load on the plate, so $F = 0$. The vertical force balance for flexure of a thin elastic plate floating on the mantle is described by the differential equation

$$D\left(\frac{\partial^4}{\partial x^4} + 2\frac{\partial^2}{\partial x^4 \partial y^2} + \frac{\partial^4}{\partial y^4}\right) w(\mathbf{x}) + (\rho_m - \rho_w)\, g w(\mathbf{x}) = -\,(\rho_c - \rho_w)\, g t_o(\mathbf{x}) \tag{17.2}$$

where the parameters are defined in Table 17.1. Note that this is the 2-D flexure equation similar to the 1-D equation (8.1) and also used in Banks et al. (1977).

Take the 2-D Fourier transform of equation (17.2) to reduce the differential equation to an algebraic equation

$$D(2\pi)^4\left(k_x^4 + 2k_x^2 k_y^2 + k_y^4\right) W(\mathbf{k}) + (\rho_m - \rho_w)\, g W(\mathbf{k})$$
$$= -\,(\rho_c - \rho_w)\, g\left[T(\mathbf{k}) - W(\mathbf{k})\right] \tag{17.3}$$

where we have used equation (17.1) to replace $T_o(\mathbf{k})$. With a little algebra and noting that $|\mathbf{k}|^4 = (k_x^2 + k_y^2)^2$, this can be rewritten as

$$D\,(2\pi\,|\mathbf{k}|)^4\, W(\mathbf{k}) + (\rho_m - \rho_c)\, g W(\mathbf{k}) = -\,(\rho_c - \rho_w)\, g T(\mathbf{k}). \tag{17.4}$$

Table 17.1.

Parameter	Definition	Value/Unit
$w(\mathbf{x})$	deflection of plate (positive up)	m
$D = \frac{Eh^3}{12(1-v^2)}$	flexural rigidity	N m
h	elastic plate thickness	m
ρ_w	seawater density	1025 kg m^{-3}
ρ_c	crust density	2800 kg m^{-3}
ρ_m	mantle density	3300 kg m^{-3}
g	acceleration of gravity	9.82 m s^{-2}
E	Young's modulus	6.5×10^{10} Pa
v	Poisson's ratio	0.25

Now one can solve for the deflection of the elastic plate in terms of the observed topography.

$$W(\mathbf{k}) = \frac{-(\rho_c - \rho_w)}{(\rho_m - \rho_c)} \left[1 + \frac{D(2\pi |\mathbf{k}|)^4}{g(\rho_m - \rho_c)} \right]^{-1} T(\mathbf{k}) \qquad (17.5)$$

This equation is called the *isostatic response function* because it describes the topography of the Moho in terms of the topography of the seafloor. Define the flexural wavelength as

$$\lambda_f = 2\pi \left[\frac{D}{g(\rho_m - \rho_c)} \right]^{1/4} = \sqrt{2}\pi\alpha. \qquad (17.6)$$

Note that α is the flexural parameter from Chapter 8. When the wavelength of the topography is much greater than the flexural wavelength, then the topography of the Moho follows the Airy-compensation model; this is *compensated* topography.

$$W(\mathbf{k}) = \frac{-(\rho_c - \rho_w)}{(\rho_m - \rho_c)} T(\mathbf{k}) \qquad (17.7)$$

In contrast, when the wavelength of the topography is much less than the flexural wavelength, the topography of the Moho is zero; this is *uncompensated* topography. The gravity field of the earth is very sensitive to the degree of compensation, so it is useful to develop the gravity field for this model.

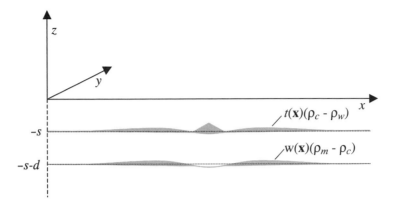

Figure 17.2

17.3 Gravity/Topography Transfer Function

The gravity anomaly for this model is approximated by compressing the topography into a sheet mass where the surface density is $(\rho_c - \rho_w) t(\mathbf{x})$. Similarly, the Moho topography is compressed into a sheet mass with surface density $(\rho_m - \rho_c) w(\mathbf{x})$. Finally, the gravity anomaly in each layer is upward-continued to the ocean surface. See Figure 17.2.

The solution to Poisson's equation (16.11) provides an approximate method of constructing a gravity model for the combined model.

$$\Delta g(\mathbf{k}) = 2\pi G \left(\rho_c - \rho_w\right) e^{-2\pi|\mathbf{k}|s} T(\mathbf{k}) + 2\pi G \left(\rho_m - \rho_c\right) e^{-2\pi|\mathbf{k}|(s+d)} W(\mathbf{k})$$
(17.8)

Using equation (17.5), this can be rewritten in terms of the observed topography.

$$\Delta g(\mathbf{k}) = 2\pi G \left(\rho_c - \rho_w\right) e^{-2\pi|\mathbf{k}|s} \left\{ 1 - \left[1 + \frac{D \left(2\pi |\mathbf{k}|\right)^4}{g \left(\rho_m - \rho_c\right)} \right]^{-1} e^{-2\pi|\mathbf{k}|d} \right\} T(\mathbf{k})$$
(17.9)

This formulation provides a direct approach to constructing gravity anomaly models from seafloor topography: i) take the 2-D Fourier transform of the topography; ii) multiply by the gravity-to-topography transfer function $Q(|\mathbf{k}|) = \Delta g/T$, and iii) take the inverse Fourier transform of the result. The most important parameter is the elastic plate thickness that is used to estimate the flexural rigidity. Figure 17.3 shows the gravity/topography transfer function for a range of elastic thicknesses. Since the asthenosphere relieves stresses on geological timescales, there is no

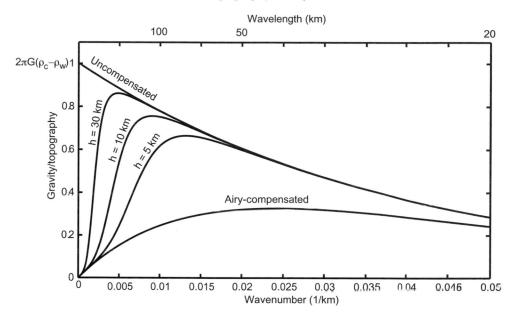

Figure 17.3

truly uncompensated topography. Thus, the gravity anomaly for very large-scale structures, such as continents and hot-spot swells, is small or zero far from the edges of these features. It is only the sharp topographic features, such as large seamounts, that will have prominent gravity expressions.

17.4 Geoid/Topography Transfer Function

Using the formulas for converting between geoid and gravity (derived in Chapter 14 on Laplace's equation) it is a simple matter to develop the geoid/topography transfer function

$$N(\mathbf{k}) = \frac{1}{2\pi |\mathbf{k}| g} \Delta g(\mathbf{k})$$

$$\frac{N(\mathbf{k})}{T(\mathbf{k})} = \frac{G(\rho_c - \rho_w)}{|\mathbf{k}| g} e^{-2\pi |\mathbf{k}| s} \left\{ 1 - \left[1 + \frac{D(2\pi |\mathbf{k}|)^4}{g(\rho_m - \rho_c)} \right]^{-1} e^{-2\pi |\mathbf{k}| d} \right\}.$$

(17.10)

This geoid topography transfer function has some interesting properties, as illustrated in Figure 17.4.

The amplitude of the geoid/topography transfer function is typically 0 to 4 meters per kilometer. This means that a seamount that is 1 km tall will produce a bump on the ocean surface that is about 1 meter tall. Since satellite altimeters have accuracy

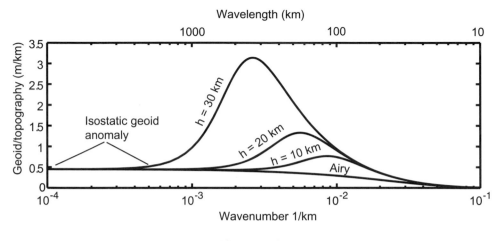

Figure 17.4

of better than 0.1 m, such seamounts will be apparent in global geoid height maps. The shape of the geoid/topography transfer functions also depends on the thickness of the elastic plate.

There is one important difference between the geoid and the gravity anomaly. At long wavelengths, the gravity/topography transfer function goes to zero, because the topography is isostatically compensated, so the gravity signal from the topography is exactly cancelled by the gravity signature from the Moho. In contrast, the geoid/topography transfer function goes to a constant value at long wavelengths. We'll exploit this long-wavelength behavior in the next section to develop a very simple approach to constructing geoid height models.

17.5 Isostatic Geoid Anomalies

The book *Geodynamics* (Section 5.12) has a nice discussion of isostatic geoid anomalies, including several of the more important applications of this approximation. However, their derivation is performed in the space domain. The wavenumber domain derivation is easier to understand and it follows the methods used in this book. In Chapter 15 on Poisson's equation, we derived the formula for computing geoid height from an arbitrary 3-D density model $\Delta\rho(x, y, z)$

$$N(\mathbf{k}) = \frac{G}{g} \int_{-\infty}^{0} \Delta\rho\,(\mathbf{k}, z) \, \frac{e^{2\pi |\mathbf{k}| z}}{|\mathbf{k}|} \, dz \qquad (17.11)$$

where $\Delta\rho(\mathbf{k}, z) = \Im^2\,[\Delta\rho(\mathbf{x}, z)]$. See Figure 17.5.

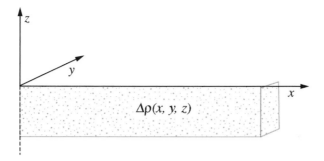

Figure 17.5

If the topography is isostatically compensated, then

$$0 = \int_{-\infty}^{0} \Delta\rho(\mathbf{x}, z)\, dz = \int_{-\infty}^{0} \Delta\rho(\mathbf{k}, z)\, dz. \tag{17.12}$$

Now expand the exponential in equation (17.11) in a Taylor series about zero wavenumber.

$$N(\mathbf{k}) = \frac{G}{g\,|\mathbf{k}|} \int_{-\infty}^{0} \Delta\rho(\mathbf{k}, z)\left[1 + 2\pi\,|\mathbf{k}|\,z + \frac{(2\pi\,|\mathbf{k}|\,z)^2}{2!} + \cdots\right] dz \tag{17.13}$$

Note that the first term in the brackets represents the integral of the density anomaly over depth, and that this is zero because of isostasy (equation (17.12)). Next, we assume that the wavelength of the anomaly is much greater than the depth of compensation $\lambda \gg z$ or $|\mathbf{k}|z \ll 1$. In this long wavelength limit, the third and higher-order terms are small compared with the second term. The integral simplifies to

$$N(\mathbf{k}) = \frac{2\pi G}{g} \int_{-\infty}^{0} \Delta\rho(\mathbf{k}, z)\, z\, dz. \tag{17.14}$$

Now take the inverse Fourier transform of equation (17.14)

$$N(\mathbf{x}) = \frac{2\pi G}{g} \int_{-\infty}^{0} \Delta\rho(\mathbf{x}, z)\, z\, dz. \tag{17.15}$$

This is a remarkable result because this formula provides a way to compute the geoid height simply by integrating the density anomaly only over depth. This integration can be done for a variety of models, including Airy compensation, Pratt compensation, and thermal compensation (i.e., spreading ridge or thermal swell). Several of these integrals are done in *Geodynamics*.

17.6 Geoid Height for Plate Cooling Model

In Chapter 5 we calculated the temperature, heat flow, and thermal subsidence for the plate cooling model. Given this simple formula (17.15) for computing the geoid for long-wavelength isostatically compensated topography and the formulas for the temperature (5.58) and seafloor depth (5.63), we can compute the decrease in geoid height versus age.

As in the depth-age calculation, we assume that temperature and density are related by the coefficient of thermal expansion as follows.

$$\rho(t,z) = \rho_m \{1 + \alpha [T_m - T(t,z)]\} \tag{17.16}$$

Since we are interested in variations in geoid height we subtract the ridge crest density-depth function from the density structure at all ages.

$$\rho(t,z) = \begin{bmatrix} (\rho_w - \rho_m) & 0 < z < d(t) \\ \alpha \rho_m [T_m - T(t, z - d(t))] & d(t) < z < L + d(t) \end{bmatrix} \tag{17.17}$$

Inserting this density/depth function into (17.15) and setting $z' = z/L$ and $d' = d/L$ we find

$$N(t) = \frac{-2\pi G L^2}{g} \left\{ \int_0^{d'(t)} (\rho_w - \rho_m) z' \, dz' - \alpha \rho_m (T_m - T_o) \right.$$

$$\left. \times \int_0^1 \left[1 - z' - \frac{2}{\pi} \sum_{n=1}^{\infty} \frac{\sin(n\pi z')}{n} e^{-\kappa(\frac{n\pi}{L})^2 t} \right] z' \, dz' \right\}. \tag{17.18}$$

Note that as in the previous integrations for depth and geoid height, we have started the thermal integration at the seafloor. The second integral is evaluated by interchanging the order of summation and integration. The interchange is valid for all $t > 0$; when $t = 0$ the series is not uniformly convergent and $N(0)$ is undefined. For greater ages, the geoid-age relation is

$$N(t) = \frac{-2\pi G L^2}{g} \left\{ (\rho_m - \rho_w) \frac{d'(t)^2}{2} \right.$$

$$\left. + \alpha \rho_m (T_m - T_o) \left[\frac{1}{6} + \frac{2}{\pi^2} \sum_{n=1}^{\infty} \frac{-1^n}{n^2} e^{-\kappa(\frac{n\pi}{L})^2 t} \right] \right\}. \tag{17.19}$$

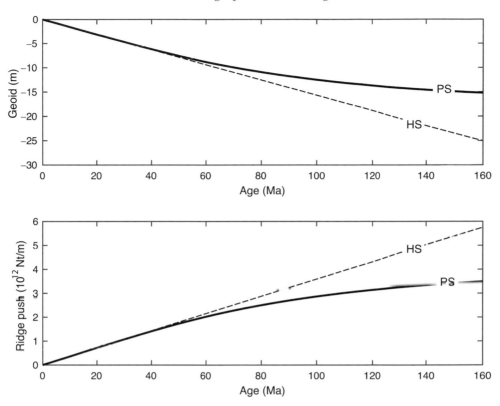

Figure 17.6 (upper) Geoid versus age for the plate (PS 125 km) and half space (HS) cooling models. (lower) Ridge push force for the plate and half space cooling models.

Figure 17.6 shows a plot of geoid height versus age for the plate cooling model (solid curve) and the half-space cooling model (dashed curve). The geoid height decreases approximately linearly with age for both models while the curve begins to flatten for the plate model at about 50 Ma. The geoid height is more sensitive than seafloor depth to the compensation in the lower lithosphere because the geoid is the integral of the density anomaly times the depth while the seafloor depth is the integrated density anomaly with no depth weighting. While this suggests that geoid height data are more important for constraining the details of lithospheric cooling at ages greater than 50 Ma, it is extremely difficult to extract a unique geoid-age signal from the Earth's geoid because the lithospheric contribution to the geoid of about 10 m is dominated by contributions from deeper sources.

There have been many studies to extract geoid height versus age using marine geoid data from satellite altimetry. As noted above, the marine geoid is dominated by

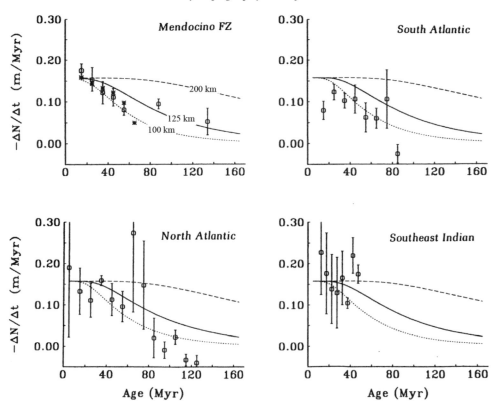

Figure 17.7 Estimates of the change in geoid with age averaged over 10 Ma age intervals for the Mendocino FZ, the North Atlantic south of 32 N, the South Atlantic, and the Southeast Indian Ridge (Sandwell and Schubert, 1982b).

long wavelength signals mostly arising from deep in the mantle. To minimize the adverse effects of the long-wavelength background signal, investigators have used shorter wavelength geoid signals across large age offset oceanic fracture zones and slower spreading ridges to estimate the change in geoid height with age offset as a function of age (e.g., Figure 17.7 (upper left)). The data are highly scattered due to the inaccurate removal of the deeper geoid signal. Nevertheless the results are consistent with a plate cooling model having an asymptotic lithospheric thickness L of between 100 and 125 km.

17.7 Isostatic Geoid and the Swell Push Force

The three driving forces of plate tectonics are slab pull, asthenospheric drag, and ridge push. Ridge push force is the outward force due to isostatically compensated topography. This mainly occurs as a gravitational sliding force on the flanks of the

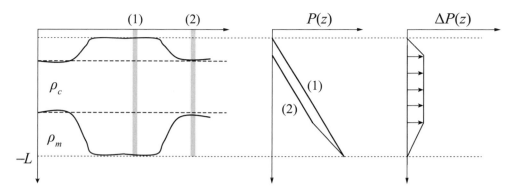

Figure 17.8

seafloor spreading ridges. However, it is also associated with the outward force due to thick continental crust or a thermal swell. One of your homework problems was to develop a general expression for this swell push force for isostatically compensated topography. Here, we'll derive this expression again and point out that the force integral is identical to the integral for computing isostatic geoid anomaly. Therefore, the two are related and one can use geoid height to measure ridge push force (Parsons and Richter, 1980; Dalen, 1981; Fleitout and Froidevaux, 1982, 1983).

Consider isostatically compensated topography as shown in Figure 17.8.

While this diagram is related to a specific Airy-type compensation mechanism, the integral relation is, in fact, quite general. To calculate the total outward force F_s due to this isostatically compensated plateau, we integrate the difference in pressure between column 1 and column 2 over depth, to the depth of compensation $-L$ (i.e., where the pressure difference is zero):

$$F_s = \int_{-L}^{0} \Delta P(z)\,\mathrm{d}z \qquad (17.20)$$

Integrate by parts:

$$F_s = \Delta P(z)z\big|_{-L}^{0} - \int_{-L}^{0} \frac{\partial \Delta P(z)}{\partial z} z\,\mathrm{d}z. \qquad (17.21)$$

Note that the first term on the right is zero because of isostasy. The second term can be written in terms of the density by noting that the vertical gradient in the pressure difference is:

$$\frac{\partial \Delta P(z)}{\partial z} = -g\Delta\rho(z). \qquad (17.22)$$

The result is

$$F_s = g \int_{-L}^{0} \Delta\rho(z)z \, dz. \qquad (17.23)$$

Comparing equation (17.15) to equation (17.23), we see the integrals are the same, so there is a direct relationship between swell push force and geoid height:

$$F_s = \frac{g^2}{2\pi G}N. \qquad (17.24)$$

This formula (17.24) can be used to calculate the ridge push force from the geoid versus age relation for the plate cooling model (17.19). Similarly, the formula (17.24) can be used to calculate the geoid versus age relation from the ridge push force of the half-space cooling model derived in Chapter 7, equation (7.16). A plot of ridge push force versus age for these two models is shown in Figure 17.7 (lower). This is the only driving force of plate tectonics that can be measured.

17.8 Exercises

Exercise 17.1 As discussed in Chapter 2, the Fourier transform of the derivative of a function is equal to $i2\pi k$ times the Fourier transform of the original function where k is the wavenumber (1/wavelength), and i is the square root of -1. Show that this relationship also holds for a discrete time series by carrying out the operations on the computer. Use the first difference formula to compute the derivative of the geoid height profile. Also compute the derivative by multiplication in the Fourier domain. Apply a phase shift to the FFT derivative so it will be aligned with the first difference derivative. Compare results. Obtain the data at: topex.ucsd.edu/pub/class/geodynamics/hw1.

Exercise 17.2 Given the relationship between gravity anomaly and topography in the Fourier transform domain provided in equation (17.9):

(a) Plot this transfer function (i.e., $Q(|\mathbf{k}|) = \Delta g/T$) for wave numbers $|\mathbf{k}|$ ranging from 0 to 10^{-3} m^{-1}. Use elastic thicknesses of 0 m and 30,000 m. Why does the transfer function approach zero at high wave numbers? Why does it approach zero at low wave numbers?

(b) Explain what happens when the elastic thickness is zero and derive the relationship between topography (above the base level) and the total crustal thickness.

(c) Using this transfer function $Q(|\mathbf{k}|)$ and the topography given in the computer file topex.ucsd.edu/pub/class/geodynamics/hw6, calculate a model gravity anomaly profile for $h = 0$. The basic procedure is to take the Fourier transform of the topography, multiply by the transfer function, and inverse Fourier transform the result.

(d) Compare this model gravity profile with the observed gravity profile. Increase the elastic thickness until the model gravity profile matches the observed gravity profile. How does this value of elastic thickness compare with the value found by Watts (1978)? How old was the lithosphere when this seamount formed?

Exercise 17.3

(a) Use the formula for the geoid height due to long wavelength isostatically compensated topography equation (17.15) to calculate the geoid height due to the following density model:

$$\Delta\rho = \sigma \left[\delta\left(z \right) - \delta\left(z + a \right) \right]$$

(b) What is the change in geoid height for a topography step of 4 km, a density of 2800 kg m^{-3}, and a compensation depth a of 40 km?
(c) What is the magnitude of the outward swell push force? What is the depth-averaged stress needed to maintain this topography?

18

Postglacial Rebound

18.1 Introduction and Dimensional Analysis

This chapter considers the classic problem of the viscous response of the mantle to rapid melting of the ice sheets following the last glacial maximum. The approach is similar to that in *Geodynamics* (Turcotte and Schubert, 2014, Section 6.10), but is for an arbitrary-shaped initial topography rather than a single wavelength cosine function. The initial condition is shown in Figure 18.1.

The main parameters are:

$$
\begin{array}{ll}
T(x) & \text{initial topography (m)} \\
\eta & \text{viscosity (Pa s)} \\
\rho & \text{density (kg m}^{-3}\text{)} \\
g & \text{acceleration of gravity (m s}^{-2}\text{)} \\
u & x\text{-velocity (m s}^{-1}\text{)} \\
w & z\text{-velocity (m s}^{-1}\text{)}
\end{array}
$$

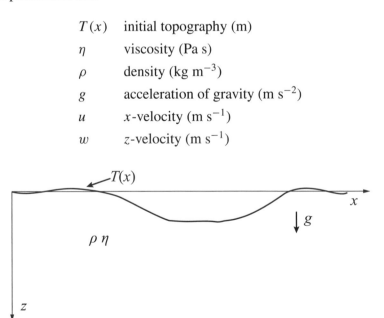

Figure 18.1 Viscous half space with an initial surface topography that decays exponentially with time under the restoring force of gravity.

Guess at a Solution: Dimensional Analysis

We can make an initial guess at the time evolution of the topography $T(t)$ assuming a single wavelength λ for the initial surface topography $T(0)$. The guess is

$$T(t) = T(0)e^{-t/\tau_r}. \tag{18.1}$$

The relaxation time should increase as the viscosity increases and decrease as the restoring force increases, so we put these in the numerator and denominator, respectively: $\eta/\rho g$. However, the result has dimensions of m s. To make this into a time we can divide by the wavelength resulting in our initial guess at the relaxation time $\tau_r = \eta/\rho g\lambda$. We'll compare this with the relaxation time based on the full derivation.

18.2 Exact Solution

We'll assume the mantle is incompressible ($\nabla \cdot \vec{u} = 0$) and the model is 2-D, so this requires that

$$\frac{\partial u}{\partial x} = -\frac{\partial w}{\partial z}. \tag{18.2}$$

As discussed in *Geodynamics*, we can define a stream function $\psi(x, z)$ such that

$$u = -\frac{\partial \psi}{\partial z}, \quad w = \frac{\partial \psi}{\partial x}. \tag{18.3}$$

This ensures that the material is incompressible.

$$\frac{\partial u}{\partial x} + \frac{\partial w}{\partial z} = -\frac{\partial^2 \psi}{\partial x \, \partial z} + \frac{\partial^2 \psi}{\partial z \, \partial x} = 0 \tag{18.4}$$

The stresses can be related to the stream function as

$$\tau_{xx} = 2\eta \frac{\partial u}{\partial x} = -2\eta \frac{\partial^2 \psi}{\partial x \, \partial z}$$

$$\tau_{zz} = 2\eta \frac{\partial w}{\partial z} = 2\eta \frac{\partial^2 \psi}{\partial z \, \partial x} \tag{18.5}$$

$$\tau_{xz} = \eta \left(\frac{\partial^2 \psi}{\partial x^2} - \frac{\partial^2 \psi}{\partial z^2} \right).$$

The force balance equations become (i.e. *Geodynamics*, equations (6.72) and (6.73)).

$$\eta \left(\frac{\partial^3 \psi}{\partial x^2 \partial z} + \frac{\partial^3 \psi}{\partial z^3} \right) + \frac{\partial P}{\partial x} = 0$$

$$\eta \left(\frac{\partial^3 \psi}{\partial x^3} + \frac{\partial^3 \psi}{\partial z^2 \partial x} \right) - \frac{\partial P}{\partial z} = 0$$

(18.6)

Following *Geodynamics*, we take the derivative of the first equation with respect to z, the second equation with respect to x, and add them to obtain the biharmonic equation

$$\frac{\partial^4 \psi}{\partial x^4} + 2 \frac{\partial^4 \psi}{\partial z^2 \partial x^2} + \frac{\partial^4 \psi}{\partial z^4} = \nabla^4 \psi = 0.$$

(18.7)

The boundary conditions for this problem are that the solution must vanish for large z, and that the surface shear stress is zero.

$$\lim_{z \to \infty} \psi(x, z) = 0$$

$$\left. \frac{\partial P}{\partial x} \right|_0 = -\rho g \frac{\partial T}{\partial x}$$

(18.8)

Note that uniform topography does not drive any flow. The flow is driven by the horizontal pressure gradient set up by the topography gradient. Take the Fourier transform of the biharmonic equation with respect to x to arrive at

$$(i 2\pi k)^4 \psi + 2(i 2\pi k)^2 \frac{\partial^2 \psi}{\partial z^2} + \frac{\partial^4 \psi}{\partial z^4} = 0.$$

(18.9)

We guess a general solution of the form

$$\psi(k, z) = A(k) e^{-2\pi |k| z} + B(k) z e^{-2\pi |k| z} + C(k) e^{2\pi |k| z} + D(k) z e^{2\pi |k| z}.$$

(18.10)

The solution must go to zero for large z, so $C = D = 0$, and the remaining terms are

$$\psi(k, z) = (A + Bz) e^{-2\pi |k| z}.$$

(18.11)

Next, we use the boundary condition that the shear traction must vanish at the surface.

$$\tau_{xz}|_0 = \eta \left(\frac{\partial^2 \psi}{\partial x^2} - \frac{\partial^2 \psi}{\partial z^2} \right) \bigg|_0 = 0$$

(18.12)

We need to compute these derivatives:

$$\frac{\partial^2 \psi}{\partial x^2} = -(2\pi k)^2 \psi$$

$$\frac{\partial \psi}{\partial z} = [B - 2\pi |k| (A + Bz)] e^{-2\pi |k| z}$$

(18.13)

and

$$\frac{\partial^2 \psi}{\partial z^2} = \left[-2B (2\pi |k|) + (2\pi |k|)^2 (A + Bz) \right] e^{-2\pi |k| z}$$

The boundary condition becomes

$$\left. \frac{\tau_{xz}}{\eta} \right|_0 = -2(2\pi |k|)^2 A + 2 (2\pi |k|) B = 0 \quad \text{so} \quad B = 2\pi |k| A.$$

(18.14)

The stream function and the two velocity components are

$$\psi (k, z) = A (1 + 2\pi |k| z) e^{-2\pi |k| z}$$

$$u = A(2\pi |k|)^2 z e^{-2\pi |k| z}$$

(18.15)

$$w = A (i 2\pi k) (1 + 2\pi |k| z) e^{-2\pi |k| z}.$$

Finally, we need to match the surface pressure gradient boundary condition

$$\left. \frac{\partial P}{\partial x} \right|_0 = -\rho g \frac{\partial T}{\partial x}.$$

(18.16)

From the force balance equation we have

$$\frac{\partial P}{\partial x} = -\eta \left(\frac{\partial^3 \psi}{\partial x^2 \partial z} + \frac{\partial^3 \psi}{\partial z^3} \right).$$

(18.17)

But we know that

$$\left. \frac{\partial \psi}{\partial z} \right|_0 = 0$$

(18.18)

so only one term remains. In the transform domain this boundary condition becomes

$$\eta \frac{\partial^3 \psi}{\partial z^3} = \rho g (i 2\pi k) T (k).$$

(18.19)

Taking the third derivative of the stream function and solving for A, we find

$$A = \frac{(i 2\pi k) \rho g T (k)}{2\eta (2\pi k)^3}.$$

(18.20)

Putting this result into the equation for the vertical velocity, we find

$$w\,(k,0) = \frac{-\rho g}{2\eta(2\pi\,|k|)}T\,(k)\,. \tag{18.21}$$

We also know that $w = \partial T/\partial t$, so we end up with a differential equation for the time evolution of the topography

$$\frac{\partial T}{\partial t} = -\frac{\rho g}{4\pi\,|k|\,\eta}T\,. \tag{18.22}$$

The solution to this differential equation is

$$T\,(k,t) = T\,(k,0)\,e^{-\frac{\rho g}{4\pi|k|\eta}t}\,. \tag{18.23}$$

From this, we find the characteristic relaxation time is

$$\tau_r = \frac{4\pi\,\eta}{\rho\,g\lambda}\,. \tag{18.24}$$

Note that this exact solution differs from the initial guess by 4π, which is OK for a first-order calculation.

Let's consider the example of Fennoscandia, which has a characteristic wavelength of 3000 km, a mantle density of 3300 kg m^{-3}, and a characteristic relaxation time of 4400 yr. Using the formula, we arrive at a viscosity of 1.1×10^{21} Pa s.

18.3 Elastic Plate over a Viscous Half Space

We can construct a more realistic model by placing an elastic lithosphere over a viscous half space. We begin with an ice load that has deformed the lithosphere for infinite time, so the deflection of the elastic plate $T\,(x)$ follows the flexural response function given in Chapter 8. The differential equation for a line load of magnitude V_o is

$$D\frac{\partial^4 T}{\partial x^4} + \rho g T\,(x) = V_o\,\delta\,(x) \tag{18.25}$$

where D is the flexural rigidity defined in Chapters 8 and 17. The solution for the deflection of the plate in the Fourier transform domain is

$$T\,(k) = \frac{V_o}{\rho g}\left[1 + \frac{k^4}{k_f^4}\right]^{-1} \tag{18.26}$$

where

$$k_f = \frac{1}{2\pi}\left[\frac{\rho g}{D}\right]^{1/4}$$

is the flexural wavenumber and is equal to the inverse flexural wavelength. At time zero, we remove this line load from the lithosphere and solve for the viscous rebound of the lithosphere and mantle. From equation (18.23), we find that the topography at some later time is

$$T(k,t) = \frac{V_o}{\rho g}\left[1 + \frac{k^4}{k_f^4}\right]^{-1} e^{-\frac{\rho g}{4\pi|k|\eta}}. \tag{18.27}$$

The response at a later time is easily computed with the following MATLAB program, where we construct the Fourier transform of the line-load flexure, multiply each wavenumber by the appropriate exponential decay, and inverse transform.

```
%
%   MATLAB program to compute rebound of an elastic plate over a viscous half space
%
    clear;clf;
    L=10000000;                  % make region wide enough to avoid edge effects
    g0=9.8;                      % acceleration of gravity
    E=7.e10;                     % Young's modulus
    h=100000;                    % elastic plate thickness
    nu=.25;                      % Poisson's ratio
    rho=3300;                    % mantle density
    eta=5.e20;                   % dynamic viscosity
    V0=4000*rho*g0;              % ice sheet load in Pascals
    D=E*h.^{3}/(12*(1-nu*nu));   % flexural rigidity
    kf=(rho*g0/D).^.25/(2.*pi);  % flexural wavenumber = 1/flexural_wavelength
    tmax=10000*86400*365;        % maximum time
%
%   set the location of the line load at L/2
%
    N=1024; dx=L/N; x=(1:N)*dx;
    topo=zeros(N,1); topo(N/2)=V0/(rho*g0);
%
%   take the Fourier transform of the load and generate wavenumbers
%
    ctopo=fftshift(fft(topo));
    k=-N/2:1:(N/2-1); k=(k./L)'; ak=abs(k);
%
%   compute the Fourier transform of the load as well as the
%   response time for each wavenumber
%
    CW0=-1./(1+(k./kf).^{4});
    TR=4*pi*eta.*abs(k)/(rho*g0);
%
%   calculate the topography at 4 time steps
%
    nstep=4; dt=tmax/nstep;
    for i = 1:nstep
    time=dt*(i);
    years=time/(86400*365)
    cmod=ctopo.*CW0.*exp(-time./TR);
    mod=real(ifft(fftshift(cmod)));
    subplot(3,1,1),plot(x/1000.,mod);axis([3000,7000,-70,40]);ylabel('topography (m) ')
    if i == 1
      hold
%
%   also plot stream function for this case
%
```

```
nd=20; dmax=2000000; dz=dmax/(nd-1);
stream=zeros(nd,N);
Ak=i*2*pi*rho*g0*k.*cmod./(2*eta*(2*pi*ak).^{3});
for n = 1:nd
  z=(n-1)*dz; argz=2*pi*ak*z;
  cstr=Ak.*(1+argz).*exp(-argz);
  str=imag(ifft(fftshift(cstr)));
  for jj = 1:N
    stream(n,jj)=str(jj);
  end
end
end
zz=-(0:nd-1)*dz;
subplot(3,1,2),contour(x/1000,zz/1000,stream,40); axis('equal');
axis([3000,7000,-1000,0]); xlabel('distance (km)'); ylabel('depth (km)');
end
```

We show the topography at four times following the removal of the load in Figure 18.2 (top). The topography rebounds in the area where the load was removed and subsides on either side of the load. The characteristic wavelength of this signature is about equal to the flexural wavelength, which is related to the thickness of the lithosphere. We found in Chapter 9 that a typical elastic thickness for flexure on million year and greater timescales is 30 km. However, to explain the width

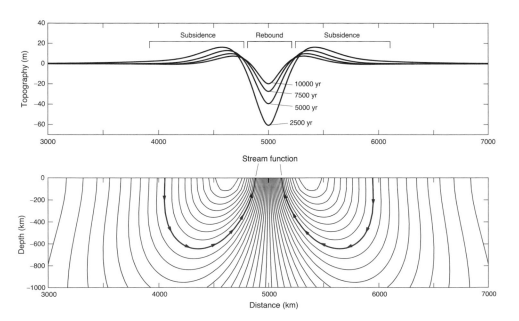

Figure 18.2 (top) Topography calculated at four times after the removal of a line load from an elastic plate over a viscous half space. (bottom) Stream function 10,000 years after the removal of the load shows upward flow of the mantle beneath the rebound area and downward flow in surrounding areas.

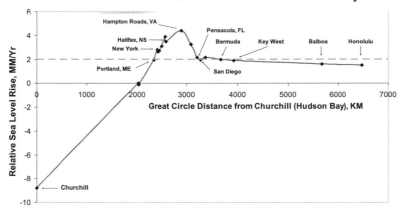

Figure 18.3 Rate of relative sea level rise versus radial distance from Hudson Bay, Canada derived from tide gauge records. The average global sea level rise over the past 100 years was about 2 mm/yr (green dashed line) although it has increased to about 2.8 mm/yr today. Areas within the perimeter of the ancient Laurentide ice sheet are still rebounding at rates up to 9 mm/yr, so relative sea level is falling. Areas on the peripheral bulge are subsiding due to unflexing of the 100 km thick elastic plate. Ocean islands far from tectonic features provide the best sites for measuring global sea level rise. From Miller and Douglas (2006).

of the peripheral bulge seen in vertical velocity profiles across Laurentide and Fennoscandia (e.g., Figure 18.3) a much thicker elastic lithosphere (~100 km) is needed.

This vertical land motion can have a significant effect on relative sea level at shorelines. The mean global sea level rise due to ocean volume changes (mass and thermal) is ~2.8 mm/yr. However, if you live in Churchill, Canada, where emergence rate is maximum, you will record a sea level fall of 9 mm/yr. If you live on the peripheral bulge in Hampton Roads, USA, you will record a sea level rise of more than 4 mm/yr.

Of course, this 2-D line load model with uniform mantle viscosity is only a crude approximation to the actual 3-D loading and viscosity structure. Moreover, the redistribution of mass as shown by the stream function in Figure 18.2 (bottom) produces a change in the geoid height (Mitrovica et al., 2009), which further complicates the global variations in sea level. Nevertheless, this simple 2-D model explains the basic features observed in past and present-day postglacial rebound.

Table 18.1. *Parameters needed for Exercise 18.1.*

Parameter	Definition	Value/Unit
κ	Thermal diffusivity	8×10^{-7} m^2 s^{-1}
h	Wine cellar depth	3 m
E	Young's modulus	6.5×10^{10} Pa
η	Dynamic viscosity	10^{20} Pa s
ρ	Mantle density	3300 kg m^{-3}
g	Acceleration of gravity	9.82 m s^{-2}
λ	Wavelength of surface deformation	3000 km

18.4 Exercises

Exercise 18.1 Given the parameters in Table 18.1, develop (guess) characteristic times for the following processes:

(a) *Heat diffusion* Describe this timescale in terms of an experiment or process.
(b) *Maxwell viscoelastic relaxation* Describe this timescale in terms of an experiment or process.
(c) *Glacial rebound, viscosity* Describe this timescale in terms of an experiment or process.

19

Driving Forces of Plate Tectonics

19.1 Introduction

The major forces acting to drive the tectonic plates are nicely presented in the classic paper by Forsyth and Uyeda (1975). They are: (1) the gravitational sliding force of the cooling oceanic lithosphere also called the ridge push force F_r; (2) the slab pull force F_s that is caused by the negative buoyancy of the cold subducted lithosphere; and (3) the viscous shear coupling τ (usually drag) on the base of the plate and both surfaces of the subducted plate (Figure 19.1). The ridge push and slab pull are body forces having units of force per length along the strike of the trench. The viscous drag is a stress having a force per area. The magnitude of the drag force depends on the ridge-to-trench plate length, the plate speed, and the viscosity. In this simple model we have an asthenosphere with a relatively low viscosity above the mesosphere having a 30–300 times higher viscosity (Hager, 1984). In addition to these main forces, there is a resistive force at the interface between the subducting oceanic plate and the overriding continental plate between the surface and depth of $\sim$40 km. This is a stick-slip zone where megathrust earthquakes are generated.

There are additional body forces (Figure 19.2) associated with phase changes within the subducting lithosphere (e.g., Arredondo and Billen, 2017). There are three major phase transitions: (1) at a depth of about 125 km, the basalt in the crust undergoes a phase change to denser eclogite; (2) at a depth of about 410 km, the primarily endothermic phase change of olivine to spinel increases the density of the cold interior of the lithosphere which adds a significant downward body force; (3) at a depth of about 660 km, the exothermic phase changes decrease the density of the cold interior of the lithosphere which decreases the downward body force and may retard the subduction through this boundary. Therefore, the total slab pull force is the sum of the thermal buoyancy force F_T and the phase change force F_p.

$$F_s = F_T + F_p \qquad (19.1)$$

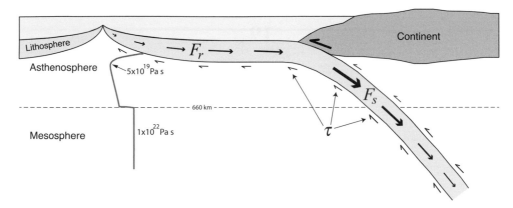

Figure 19.1 Primary thermomechanical driving forces of the tectonic plates. Viscosity versus depth (red curve) has a broad minimum defining the asthenosphere. The mesosphere begins at a depth of about 410 km and has a sharp viscosity increase at a depth of 660 km.

In the remainder of this chapter we highlight the relative importance of the thermal and phase change buoyancy forces for subducting lithosphere. In Chapter 5 we calculated the temperature and buoyancy of the lithosphere using a plate cooling model, which includes heat flow into the base of the lithosphere that limits the asymptotic plate thickness to about 125 km. A back of the envelope calculation suggests that if the plate cools for 80 Ma prior to subduction, it will take about 80 Ma to reheat after subduction. Given a typical vertical subduction rate of 30 mm/yr it will take 22 Ma to reach the 660 depth so the slab remains relatively cold and dense during its transit through the upper mantle.

19.2 Age of Subducting Lithosphere

To better understand the forces driving the plates we first examine the area versus age distribution of oceanic lithosphere (Figure 19.3) derived from a global age grid (Müller et al., 1997). The area of seafloor decreases approximately linearly with age out to 180 Ma. This suggests that all ages of lithosphere are equally likely to subduct. In Section 5.6 we computed the buoyancy of the lithosphere relative to the mantle, including the positive buoyancy of the crust and depleted upper mantle. Oceanic lithosphere having normal crustal thickness of 6–7 km remains positively buoyant until it has cooled for ∼30 Ma. However, we see from this Figure 19.3, as well as from tectonic maps, that young lithosphere is indeed subducting. There are two reasons young plates subduct: (1) Young lithosphere is physically connected to older lithosphere and thus it is being pulled into the trench by the negative buoyancy of the older plate. (2) Once the crust and depleted mantle

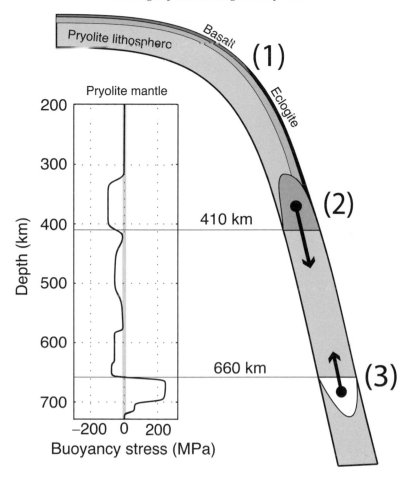

Figure 19.2 Subduction driving forces due to phase changes having the thermal structure of the plate cooling model at an age of 80 Ma. The overall buoyancy stress is the thickness-averaged upward buoyancy due to composition and phase. There are three major components: (1) As discussed in Section 5.5, pyrolite upwelling beneath ridges undergoes decompression melting at a depth of about 40 km. This melt migrates to the magma chamber at the ridge axis, where it forms oceanic crust (red) with a normal thickness of 7 km and an average density of 2900 kg m^{-3}. In addition, ultramafic residues formed by partial melting during the generation of basalt are less dense than undepleted mantle. This layer of depleted mantle (green) has a thickness of about 21 km and an average density of 3235 kg m^{-3}, which is less than the normal density of pyrolite mantle of 3300 kg m^{-3}. This adds an additional positive buoyancy to the oceanic lithosphere. During subduction, the basalt undergoes a phase change to the higher density eclogite 3540 kg m^{-3}. This dramatic increase in density of the crust cancels the overall positive buoyancy of the ultramafic residues so starting at a depth of 200 km, the net buoyancy stress is zero. (2) Endothermic phase changes (positive Clapeyron slope) produce a zone of increased density in the cold lithosphere for a large part of the transition zone between depths of 310 km and 660 km. (3) Exothermic phase changes (negative Clapeyron slope) below 660 km result in a zone of decreased density between depths of 660 and 720 km.

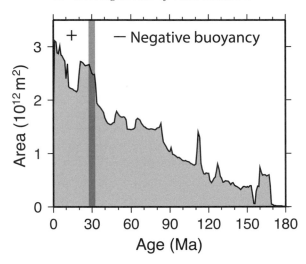

Figure 19.3 Area versus age distribution of oceanic lithosphere calculated from
the Müller et al. (1997) age map. The area decreases almost linearly with age
to the oldest seafloor at ∼180 Ma. Ocean lithosphere (6 km crust) is positively
buoyant for ages less than 30 Ma (see Section 5.6)

reach a depth of ∼125 km the basalt undergoes a phase change to denser eclogite
and the compositional buoyancy is negated (Arredondo and Billen, 2017) so the
negative thermal buoyancy dominates.

19.3 Forces due to Phase Changes

To calculate the buoyancy force due to phase changes F_p, we numerically integrate
the buoyancy stress (Arredondo and Billen, 2017) (Figure 19.2) from the surface
to the downdip end of the slab. If the subduction zone is young and the slab
penetrates to less than ∼320 km then the contribution from the phase change is
zero (Figure 19.4). If the end of the slab is at a depth of 660 km, the phase change
contribution is maximum at -18×10^{12} N m^{-1}. Most older slabs penetrate into
the lower mantle so the 660 phase change, having a positive Clapeyron slope,
dramatically reduces the F_p to -8×10^{12} N m^{-1} at a depth of 720 km. This
reduction in negative buoyancy combined with a greater than 30 times increase in
mantle viscosity is responsible for a tendency for subducted lithosphere to stagnate
at a depth of about 700 km.

19.4 Forces due to Thermal Buoyancy

Next we consider the two driving forces related to lithospheric cooling. The
ridge-push force for the plate cooling model was already indirectly computed in

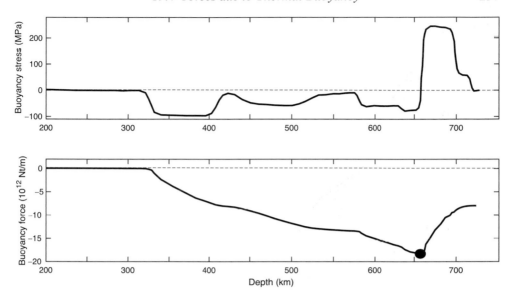

Figure 19.4 (upper) Buoyancy stress due to phase changes in lithosphere that has cooled for 80 Ma (Arredondo and Billen, 2017). (lower) Depth-integrated buoyancy force F_p in N m^{-1} (negative is downward force).

Chapter 17. The ridge push force F_r for long-wavelength, isostatically compensated topography is given by

$$F_r = \frac{g^2}{2\pi G} N \qquad (19.2)$$

where N is the geoid height, g is the acceleration of gravity, and G is the gravitational constant. We can apply this to the geoid height for the plate cooling model (equation (17.19)) so the ridge push force is

$$F_r(t) = g L^2 \left\{ (\rho_m - \rho_w) \frac{d'(t)^2}{2} + \alpha \rho_m (T_m - T_o) \left[\frac{1}{6} + \frac{2}{\pi^2} \sum_{n=1}^{\infty} \frac{-1^n}{n^2} e^{-\kappa \left(\frac{n\pi}{L} \right)^2 t} \right] \right\}.$$

$$(19.3)$$

To compute the thermal buoyancy of the slab, first assume that there is no reheating of the slab after subduction so it maintains the thermal profile of 80 Ma lithosphere. Actually, the reheating of the subducted plate transfers the negative buoyancy from the plate to the surrounding mantle so it is still available to drive the subduction. The buoyancy stress is the integral of the density anomaly relative to a uniform mantle density times the acceleration of gravity.

$$\sigma_T = g\rho_m \alpha \int_0^L (T - T_m)\, dz \tag{19.4}$$

We have done this integral in Chapter 5 and the result is

$$\sigma_T(t) = -g\alpha\rho_m (T_m - T_o) L \left[\frac{1}{2} - \frac{4}{\pi^2} \sum_{n=1}^{\infty} \frac{1}{(2n-1)^2} e^{-\kappa\left(\frac{(2n-1)\pi}{L}\right)^2 t} \right]. \tag{19.5}$$

We can integrate over the length of the subducted lithosphere to obtain the thermal component of the slab pull force versus cooling time t and slab penetration depth Z

$$F_T(t, Z) = -g\alpha\rho_m (T_m - T_o) L Z \left[\frac{1}{2} - \frac{4}{\pi^2} \sum_{n=1}^{\infty} \frac{1}{(2n-1)^2} e^{-\kappa\left(\frac{(2n-1)\pi}{L}\right)^2 t} \right]. \tag{19.6}$$

The results for the ridge push and thermal buoyancy of the slab are shown in Figure 19.5. The horizontally directed ridge-push force is 2.3×10^{12} N m^{-1} at 80 Ma. The thermal buoyancy of the slab depends on both the age of the lithosphere being subducted and the depth of penetration of the slab. This calculation also assumes the slab remains intact from the surface to the slab penetration depth so it can serve as a stress guide (Conrad and Lithgow-Bertelloni, 2002). For example, subduction of 80 Ma lithosphere to a depth of 660 km results in a thermal buoyancy of -35×10^{12} N m^{-1}. This is more than 10 times greater than the ridge push force.

19.5 Asthenospheric Drag Force

The final force that we consider is the asthenospheric drag force. Our very simple model consists of a plate of length L (10,000 km) moving at a velocity u_o (40 mm/yr) over an asthenosphere of dynamic viscosity η (5×10^{19} Pa s). The thickness of the asthenosphere is h (100 km) and there is no motion at the base of the asthenosphere. The solution to this simple channel flow model is discussed in *Geodynamics* (Section 6.2). The velocity as a function of depth is $u(z) = u_o (1 - z/h)$. The shear stress at the base of the lithosphere is $\tau = \eta \frac{\partial u}{\partial z} = -\eta u_o / h$. The total drag force F_d at a distance L from the spreading ridge is simply the stress times distance or $F_d = -\eta u_o L / h$. Using the values above we find the force is -0.6×10^{12} N m^{-1}. Note that this is much smaller than the other driving forces. This drag force could be comparable to the ridge push force if the viscosity was 5 times larger.

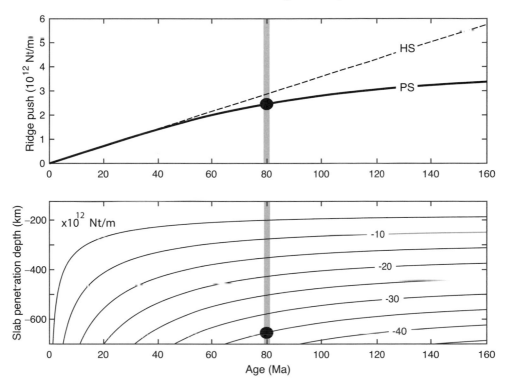

Figure 19.5 (upper) Ridge push force for plate (PS) and half-space (HS) cooling models. (lower) Contour plot of the thermal component of slab pull versus the age of the lithosphere being subducted and the depth of penetration of the end of the slab.

19.6 Discussion: Relative Magnitudes of Forces

In summary, we have decomposed the driving forces of plate motion into:

- the slab pull force with a thermal and phase change component $F_s = F_T + F_p$;
- the ridge push force F_r;
- and the drag force F_d.

Table 19.1 provides some end member estimates of the magnitude of these forces for a small young plate and a large old plate.

Note that the thermal component of the slab pull force dominates. The phase change component of slab pull is about half the thermal component. The ridge push is 10–20 times smaller than the slab pull force. If the slabs remain as coherent stress

Table 19.1. *Strength of driving forces* $\times 10^{12}$ $N\,m^{-1}$. *The sum of forward driving forces is* $F_{tot} = F_T + F_p - F_r$.

t(Ma)	Z(km)	L(km)	F_T	F_p	F_r	F_d	F_{tot}	Plate characteristic
20	300	2000	−6.5	0	0.8	−0.12	−7.3	small young plate
—	660	—	−16	−9	0.8	−0.12	−25.8	—
—	700	—	−20	−4	0.8	−0.12	−24.8	—
80	300	10000	−13	0	2.3	−0.6	−15.3	large old plate
—	660	—	−35	−18	2.3	−0.6	−55.3	—
—	700	—	−37	−8	2.3	−0.6	−47.3	—

guides they must be able to transmit stress of up to 500 MPa when the force is averaged over the 100 km thickness of the subducted lithosphere. As noted by Conrad and Lithgow-Bertelloni (2002), this is close to the maximum strength of cold lithosphere (Kohlstedt et al., 1995).

These simple calculations provide estimates of the forces driving the plates but one must develop a full numerical model of viscous flow in a layered Earth to relate these forces to the surface plate velocities and the global variations in geoid height. The simplest models require a more than 30 times increase in viscosity between the upper and lower mantle (Hager, 1984). More complex models that also match the observed speeds and directions of the plates suggest that the entire negative buoyancy of the slabs in the upper mantle is effectively transmitted to the surface to drive plate motions. These models also predict that the negative buoyancy of the slabs in the lower mantle drives deep mantle flow but at a much slower rate due to the higher viscosity (Kohlstedt et al., 1995). An example of a numerical simulation of a subducting plate is shown in Figure 19.6. This 80 Ma lithosphere has subducted for 40 Ma so a large part of the slab extends into the much higher viscosity lower mantle where if folds due to the resistive viscous drag.

19.7 Exercises

Exercise 19.1 Write a program to calculate the ridge push force versus cooling time on Venus for asymptotic plate thicknesses of 125 and 500 km. Use a surface temperature of 455 °C and a deep mantle temperature somewhat higher than the Earth at 1400 °C. How does this compare with Earth?

Exercise 19.2 Write a short essay on the arguments for and against lithospheric subduction on Venus. Begin with a brief summary of the main observations that were used to confirm lithospheric subduction on the Earth. Then discuss some of the published arguments for and against subduction on Venus. End with a discussion

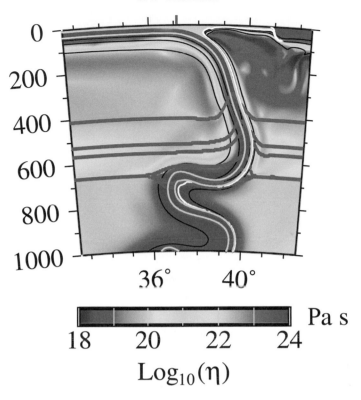

Figure 19.6 Numerical model of lithospheric subduction (Arredondo and Billen, 2017). Buoyancy due to phase changes (Figure 19.3) are combined with the thermal buoyancy to drive subduction. The 80 Ma lithosphere has a very high yield strength of 1000 MPa that enables the plate to remain intact. The upper mantle viscosity is 10^{19} Pa s while the lower mantle has a much higher viscosity of 10^{22} Pa s.

of how to prove or disprove the subduction theory with another mission to Venus. There is an extensive reference list at the following web site `topex.ucsd.edu/venus/index.html`.

Exercise 19.3 Describe the three major phase changes that occur within subducting lithosphere. Explain how they affect lithospheric buoyancy. What is the Clapeyron slope?

Bibliography

Arredondo, K., and M. Billen, Coupled effects of phase transitions and rheology in 2-d dynamical models of subduction, *Journal of Geophysical Research: Solid Earth*, *122*(7), 5813–5830, 2017.

Banks, R., R. Parker, and S. Heustis, Isostatic compensation on a continental scale: Local versus regional mechanisms, *Geophysical Journal of the Royal Astronomical Society*, *51*, 431–452, 1977.

Becker, J., et al., Global bathymetry and elevation data at 30 arc seconds resolution: SRTM30-PLUS, *Marine Geodesy*, *32*(4), 355–371, 2009.

Blakely, J., *Potential Theory in Gravity and Magnetics*, Cambridge University Press, New York, 1995.

Brace, W., and D. Kohlstedt, Limits on lithospheric stress imposed by laboratory experiments, *Journal of Geophysical Research*, *85*(B11), 6248–6252, 1980.

Bracewell, R., *The Fourier Transform and Its Applications*, second ed., McGraw-Hill, New York, 1978.

Brown, C., and R. Phillips, Flexural rift flank uplift at the Rio Grande rift, New Mexico, *Tectonics*, *18*(6), 1275–1291, 1999.

Burridge, R., and L. Knopoff, Body force equivalents for seismic dislocations, *Bulletin of the Seismological Society of America*, *54*(6A), 1875–1888, 1964.

Byerlee, J., Friction of rocks, in *Rock Friction and Earthquake Prediction*, pp. 615–626, Springer, 1978.

Caldwell, J., W. Haxby, D. E. Karig, and D. Turcotte, On the applicability of a universal elastic trench profile, *Earth and Planetary Science Letters*, *31*(2), 239–246, 1976.

Cande, S., J. LaBrecque, R. Pitman, X. Golovchenko, and W. Haxby, *Magnetic Lineations of the World's Ocean Basins*, American Association of Petroleum Geologists, Tulsa, OK, 1989.

Carslaw, H., and J. Jaeger, *Conduction of Heat in Solids*, second ed., Oxford University Press, Oxford, 1959.

Chen, Y. J., Oceanic crustal thickness versus spreading rate, *Geophysical Research Letters*, *19*(8), 753–756, 1992.

Cohen, S., Numerical models of crustal deformation in seismic zones, *Advances in Geophysics*, *41*, 134–231, 1999.

Cohen, S., and D. Smith, Lageos scientific results: Introduction, *Journal of Geophysical Research: Solid Earth*, *90*(B11), 9217–9220, 1985.

Conrad, C., and C. Lithgow-Bertelloni, How mantle slabs drive plate tectonics, *Science*, *298*(5591), 207–209, 2002.

Crough, S., Hotspot swells, *Annual Review of Earth and Planetary Sciences*, *11*(1), 165–193, 1983.

Cuffey, K., and W. Patterson, *The Physics of Glaciers*, third ed., Butterworth-Heinemann, imprint of Elsevier, Burlington, MA, 2010.

Dalen, F., Isostasy and the ambient state of stress in the oceanic lithosphere, *Journal of Geophysical Research*, *86*, 7801–7807, 1981.

DeMets, C., R. Gordon, D. Argus, and S. Stein, Current plate motions, *Geophysical Journal of the Royal Astronomical Society*, *101*, 425–478, 1990.

DeMets, C., R. Gordon, D. Argus, and S. Stein, Geologically current plate motions, *Geophysical Journal International*, *181*, 1–80, 2010.

Doin, M., and L. Fleitout, Thermal evolution of the oceanic lithosphere: An alternate view, *Earth and Planetary Science Letters*, *142*, 121–136, 1996.

Dorman, L., and B. Lewis, Experimental isostasy 3: Inversion of the isostatic Green's function and lateral density changes, *Journal of Geophysical Research*, *77*, 3068–3077, 1972.

Engdahl, E., R. van der Hilst, and R. Buland, Global teleseismic earthquake relocation with improved travel times and procedures for depth determination, *Bulletin of the Seismological Society of America*, *88*, 722–743, 1998.

Fialko, Y., Temperature fields generated by the elastodynamic propagation of shear cracks in the earth, *Journal of Geophysical Research: Solid Earth*, *109*(B1), 2004.

Fleitout, L., and C. Froidevaux, Tectonics and topography for a lithosphere containing density heterogeneities, *Tectonics*, *1*, 21–56, 1982.

Fleitout, L., and C. Froidevaux, Tectonic stresses in the lithosphere, *Tectonics*, *2*, 315–324, 1983.

Flesch, L., A. Haines, and W. Holt, Dynamics of the India-Eurasia collision zone, *Journal of Geophysical Research: Solid Earth*, *106*(B8), 16,435–16,460, 2001.

Forsyth, D., and S. Uyeda, On the relative importance of the driving forces of plate motion, *Geophysical Journal International*, *43*(1), 163–200, 1975.

Fowler, C., *The Solid Earth: An Introduction to Global Geophysics*, Cambridge University Press, Cambridge, 1990.

Garcia, E., D. Sandwell, and K. Luttrell, An iterative spectral solution method for thin elastic plate flexure with variable rigidity, *Geophysical Journal International*, *200*(2), 1012–1028, 2014.

Garcia, E., D. Sandwell, and D. Bassett, Outer trench slope flexure and faulting at Pacific basin subduction zones, *Geophysical Journal International*, *218*(1), 708–728, 2019.

Garland, G., *The Earth's Shape and Gravity*, Pergamon Press, 1977.

Gee, J., and D. Kent, Variation in layer 2A thickness and the origin of the central anomaly magnetic high, *Geophysical Research Letters*, *21*(4), 297–300, 1994.

Gee, J., and D. Kent, Source of oceanic magnetic anomalies and the geomagnetic polarity timescale, in *Treatise on Geophysics*, edited by G. Schubert, pp. 455–507, Elsevier, Amsterdam, 2007.

Goetze, C., and B. Evans, Stress and temperature in the bending lithosphere as constrained by experimental rock mechanics, *Geophysical Journal International*, *59*(3), 463–478, 1979.

Greenberg, M., *Applications of Green's Functions in Science and Engineering*, Dover Publications, 2015.

Hager, B., Subducted slabs and the geoid: Constraints on mantle rheology and flow, *Journal of Geophysical Research: Solid Earth*, *89*(B7), 6003–6015, 1984.

Hasterok, D., A heat flow based cooling model for tectonic plates, *Earth and Planetary Science Letters*, *361*, 34–43, 2013.

Haxby, W., and D. Turcotte, On isostatic geoid anomalies, *Journal of Geophysical Research*, *83*, 5, 473–5, 478, 1978.

Haxby, W., G. Garner, J. LaBrecque, and J. Weissel, Digital images of combined oceanic and continental data sets and their use in tectonic studies, *EOS Transactions American Geophysical Union*, *64*, 995–1004, 1983.

Horner-Johnson, B. C., and R. Gordon, Equatorial Pacific magnetic anomalies identified from vector aeromagnetic data, *Geophysical Journal International*, *155*(2), 547–556, 2003.

Hwang, C., and B. Parsons, An optimal procedure for deriving marine gravity from multi-satellite altimetry, *Geophysical Journal International*, *125*, 705–719, 1996.

Jackson, J., *Classical Electrodynamics*, third ed., John Wiley & Sons, New York, 1998.

Jaeger, J., N. Cook, and R. Zimmerman, *Fundamentals of Rock Mechanics*, John Wiley & Sons, 2009.

Johnson, C., and D. Sandwell, Lithospheric flexure on Venus, *Geophysical Journal International*, *119*(2), 627–647, 1994.

Kent, G., A. Harding, and J. Orcutt, Distribution of magma beneath the East Pacific rise between the Clipperton Transform and the 9° 17' deval from forward modeling of common depth point data, *Journal of Geophysical Research*, *98*, 13,945–13,969, 1993.

Kohlstedt, D., B. Evans, and S. Mackwell, Strength of the lithosphere: Constraints imposed by laboratory experiments, *Journal of Geophysical Research: Solid Earth*, *100*(B9), 17,587–17,602, 1995.

Lachenbruch, A., and J. Sass, Heat flow and energetics of the San Andreas fault zone, *Journal of Geophysical Research*, *85*, 6185–6222, 1980.

Laske, G., G. Masters, Z. Ma, and M. Pasyanos, Update on CRUST1.0 A 1-degree global model of Earth's crust, in *Geophysical Research Abstracts*, vol. 15, p. 2658, EGU General Assembly Vienna, Austria, 2013.

Leeds, A., and E. Kausel, Variations of upper mantle structure under the Pacific Ocean, *Science*, *186*, 141–143, 1974.

Levitt, D., and D. Sandwell, Lithospheric bending at subduction zones based on depth soundings and satellite gravity, *Journal of Geophysical Research: Solid Earth*, *100*, 379–400, 1995.

Luttrell, K., D. Sandwell, B. Smith-Konter, B. Bills, and Y. Bock, Modulation of the earthquake cycle at the southern San Andreas fault by lake loading, *Journal of Geophysical Research: Solid Earth*, *112*(B8), 2007.

Massell, C., Large scale structural variation of trench outer slopes and rises, Ph.D. dissertation, Scripps Institution of Oceanography, La Jolla, CA, 2002.

McKenzie, D., and C. Bowin, The relationship between bathymetry and gravity in the Atlantic Ocean, *Journal of Geophysical Research*, *81*, 1903–1915, 1976.

McNutt, M., Compensation of oceanic topography: An application of the response function technique to the Surveyor area, *Journal of Geophysical Research*, *84*, 7,589–7,598, 1979.

McNutt, M., and H. Menard, Lithospheric flexure and uplifted atolls, *Journal of Geophysical Research: Solid Earth*, *83*(B3), 1206–1212, 1978.

McNutt, M., and H. Menard, Constraints on yield strength in the oceanic lithosphere derived from observations of flexure, *Geophysical Journal of the Royal Astronomical Society*, *71*, 363–394, 1982.

Menard, H., *Marine geology of the Pacific*, McGraw-Hill, 1964.

Miller, L., and B. C. Douglas, On the rate and causes of twentieth century sea-level rise, *Philosophical Transactions of the Royal Society of London A: Mathematical, Physical and Engineering Sciences, 364*(1841), 805–820, 2006.

Minster, J., and T. Jordan, Present-day plate motions, *Journal of Geophysical Research, 85,* 5331–5354, 1978.

Mitrovica, J. X., N. Gomez, and P. U. Clark, The sea-level fingerprint of West Antarctic collapse, *Science, 323*(5915), 753, 2009.

Mueller, S., and R. J. Phillips, On the reliability of lithospheric constraints derived from models of outer-rise flexure, *Geophysical Journal International, 123*(3), 887–902, 1995.

Müller, R., W. Roest, J. Royer, L. Gahagan, et al., Digital isochrons of the world's ocean floor, *Journal of Geophysical Research, 102,* 3211–3214, 1997.

Oxburgh, E., and E. Parmentier, Compositional and density stratification in oceanic lithosphere-causes and consequences, *Journal of the Geological Society, 133*(4), 343–355, 1977.

Parker, R., The rapid calculation of potential anomalies, *Geophysical Journal of the Royal Astronomical Society, 31,* 441–455, 1973.

Parsons, B., and F. Richter, A relationship between the driving force and geoid anomaly associated with mid-ocean ridges, *Earth and Planetary Science Letters, 51,* 445–450, 1980.

Parsons, B., and J. Sclater, An analysis of the variation of the ocean floor bathymetry and heat flow with age, *Journal of Geophysical Research, 82,* 803–827, 1977.

Pavlis, N., S. Holmes, S. Kenyon, and J. Factor, The development and evaluation of the Earth Gravitational Model 2008 (EGM2008), *Journal of Geophysical Research, 117*(B4), 2012.

Priestley, K., D. McKenzie, and T. Ho, A lithosphere–asthenosphere boundary-a global model derived from multimode surface-wave tomography and petrology, *Lithospheric Discontinuities*, pp. 111–123, 2018.

Rapp, R., and Y. Yi, Role of ocean variability and dynamic topography in the recovery of the mean sea surface and gravity anomalies from satellite altimeter data., *Journal of Geodesy, 71,* 617–629, 1997.

Renkin, M., and J. Sclater, Depth and age in the North Pacific, *Journal of Geophysical Research-Solid Earth, 93,* 2919–2935, 1988.

Ryan, M. (Ed.), *Magmatic Systems*, Academic Press, San Diego, 1994.

Sandwell, D., Thermal isostasy: Response of a moving lithosphere to a distributed heat source, *Journal of Geophysical Research: Solid Earth, 87*(B2), 1001–1014, 1982.

Sandwell, D., Thermomechanical evolution of oceanic fracture zones, *Journal of Geophysical Research, 89,* 11,401–11,413, 1984.

Sandwell, D., Biharmonic spline interpolation of geos-3 and seasat altimeter data, *Geophysical Research Letters, 14*(2), 139–142, 1987.

Sandwell, D., and G. Schubert, Lithospheric flexure at fracture zones, *Journal of Geophysical Research, 87,* 4657–4667, 1982a.

Sandwell, D., and G. Schubert, Evidence for retrograde lithospheric subduction on Venus, *Science, 257*(5071), 766–770, 1992.

Sandwell, D., and W. Smith, Marine gravity anomaly from Geosat and ERS-1 satellite altimetry, *Journal of Geophysical Research, 102,* 10,039–10,054, 1997.

Sandwell, D., R. D. Müller, W. Smith, E. Garcia, and R. Francis, New global marine gravity model from CryoSat-2 and Jason-1 reveals buried tectonic structure, *Science, 346*(6205), 65–67, 2014.

Sandwell, D., H. Harper, B. Tozer, and W. Smith, Gravity field recovery from geodetic altimeter missions, *Advances in Space Research*, 2019.

Sandwell, D. T., and G. Schubert, Geoid height-age relation from seasat altimeter profiles across the Mendocino fracture zone, *Journal of Geophysical Research*, *87*(B5), 3949–3958, 1982b.

Savage, J., Equivalent strike-slip cycles in half-space and lithosphere-asthenosphere earth models, *Journal of Geophysical Research*, *95*, 4873–4879, 1990.

Savage, J., and R. Burford, Geodetic determination of relative plate motion in central California, *Journal of Geophysical Research*, *78*(5), 832–845, 1973.

Scambos, T., O. Sergienko, A. Sargent, D. MacAyeal, and J. Fastook, Icesat profiles of tabular iceberg margins and iceberg breakup at low latitudes, *Geophysical Research Letters*, *32*(23), 2005.

Schouten, H., and K. McCamy, Filtering marine magnetic anomalies, *Journal of Geophysical Research*, *77*, 7089–7099, 1972.

Schouten, J. A., A fundamental analysis of magnetic anomalies over oceanic ridges, *Marine Geophysical Researches*, *1*(2), 111–144, 1971.

Schubert, G., and D. Sandwell, Crustal volumes of the continents and of oceanic and continental submarine plateaus, *Earth and Planetary Science Letters*, *92*(2), 234–246, 1989.

Schubert, G., and D. Sandwell, A global survey of possible subduction sites on Venus, *Icarus*, *117*(1), 173–196, 1995.

Sclater, J., C. Jaupart, and D. Galson, The heat flow through oceanic and continental crust and the heat loss of the Earth, *Reviews of Geophysics and Space Physics*, *18*, 269–311, 1980.

Segall, P., *Earthquake and volcano deformation*, Princeton University Press, 2010.

Seton, M., R. D. Müller, S. Zahirovic, S. Williams, N. M. Wright, J. Cannon, J. M. Whittaker, K. J. Matthews, and R. McGirr, A global data set of present-day oceanic crustal age and seafloor spreading parameters, *Geochemistry, Geophysics, Geosystems*, *21*(10), e2020GC009,214, 2020.

Siebert, L., and T. Simkin, *Volcanoes of the World: an Illustrated Catalog of Holocene Volcanoes and their Eruptions*, vol. GVP-3, Smithsonian Institution, Global Volcanism Program, Digital Information Series, http://www.volcano.si.edu/world/, 2002.

Smith, B., and D. Sandwell, Coulomb stress accumulation along the San Andreas fault system, *Journal of Geophysical Research: Solid Earth*, *108*(B6), 2003.

Smith, B., and D. Sandwell, A three-dimensional semianalytic viscoelastic model for time-dependent analyses of the earthquake cycle, *Journal of Geophysical Research: Solid Earth*, *109*(B12), 2004.

Smith, G., and S. Banerjee, Magnetic structure of the upper kilometer of the marine crust at DSDP hole 504B, Eastern Pacific Ocean, *Journal of Geophysical Research*, *91*, 10,337–10,354, 1986.

Smith, W., and D. Sandwell, Bathymetric prediction from dense satellite altimetry and sparse shipboard bathymetry, *Journal of Geophysical Research*, *99*, 21,803–21,824, 1994.

Smith, W., and D. Sandwell, Global sea floor topography from satellite altimetry and ship depth soundings, *Science*, *277*, 1956–1962, 1997.

Smith-Konter, B., and D. Sandwell, Stress evolution of the San Andreas fault system: Recurrence interval versus locking depth, *Geophysical Research Letters*, *36*(13), 2009.

Stacey, F., *Physics of the Earth*, John Wiley & Sons, 1977.

Staudigel, H., A. Koppers, J. Lavelle, T. Pitcher, and T. Shank, Defining the word "Seamount", *Oceanography*, *23*(1), 2010.

Stein, C. A., and S. Stein, A model for the global variation in oceanic depth and heat flow with lithospheric age, *Nature*, *359*(6391), 123–129, 1992.

Steketee, J., On Volterra's dislocations in a semi-infinite elastic medium, *Canadian Journal of Physics*, *36*(2), 192–205, 1958.

Tozer, B., D. Sandwell, W. Smith, C. Olson, J. Beale, and P. Wessel, Global bathymetry and topography at 15 arc sec: Srtm15+, *Earth and Space Science*, *6*(10), 1847–1864, 2019.

Turcotte, D., and E. Oxburgh, Finite amplitude convection cells and continental drift, *Journal of Fluid Mechanics*, *28*, 29–42, 1967.

Turcotte, D., and G. Schubert, *Geodynamics*, third ed., Cambridge University Press, Cambridge, 2014.

Van Avendonk, H. J., J. K. Davis, J. L. Harding, and L. A. Lawver, Decrease in oceanic crustal thickness since the breakup of Pangaea, *Nature Geoscience*, *10*(1), 58–61, 2017.

Vaughan, D., Tidal flexure at ice shelf margins, *Journal of Geophysical Research: Solid Earth*, *100*(B4), 6213–6224, 1995.

Vink, G., W. Morgan, and W. L. Zhao, Preferential rifting of continents: a source of displaced terranes, *Journal of Geophysical Research*, *89*(B12), 10,072–10,076, 1984.

Watts, A., An analysis of isostasy in the world's oceans 1. Hawaiian-Emperor seamount chain, *Journal of Geophysical Research*, *83*(B12), 5989–6004, 1978.

Watts, A., *Isostasy and Flexure of the Lithosphere*, Cambridge University Press, 2001.

Wdowinski, S., B. Smith-Konter, Y. Bock, and D. Sandwell, Diffuse interseismic deformation across the Pacific–North America plate boundary, *Geology*, *35*(4), 311–314, 2007.

Weertman, J., and J. R. Weertman, *Elementary dislocation theory*, Macmillan, 1966.

Weissel, J., and G. Karner, Flexural uplift of rift flanks due to mechanical unloading of the lithosphere during extension, *Journal of Geophysical Research: Solid Earth*, *94*(B10), 13,919–13,950, 1989.

Wessel, P., Global distribution of seamounts inferred from gridded Geosat/ERS-1 altimetry, *Journal of Geophysical Research: Solid Earth*, *106*(B9), 19,431–19,441, 2001.

Wessel, P., and D. Bercovici, Interpolation with splines in tension: a Green's function approach, *Mathematical Geology*, *30*(1), 77–93, 1998.

Wessel, P., and W. Smith, New version of the Generic Mapping Tools released, *EOS Trans. AGU*, *76*, 329, http://gmt.soest.hawaii.edu/, 1995.

Wessel, P., J. Luis, L. Uieda, R. Scharroo, F. Wobbe, W. Smith, and D. Tian, The generic mapping tools version 6, *Geochemistry, Geophysics, Geosystems*, 2019.

Index